HIBERNATION

■ Clive Roots

Greenwood Guides to the Animal World

GREENWOOD PRESS
Westport, Connecticut · London

Library of Congress Cataloging-in-Publication Data

Roots, Clive, 1935–
    Hibernation / Clive Roots.
        p. cm.—(Greenwood guides to the animal world, ISSN 1559–5617)
    Includes bibliographical references and index.
    ISBN 0–313–33544–3 (alk. paper)
    1. Hibernation.   I. Title.   II. Series.
    QL755.R66   2006
    591.56′5—dc22   2006010621

British Library Cataloguing in Publication Data is available.

Library of Congress Catalog Card Number: 2006010621
ISBN: 0–313–33544–3
ISSN: 1559–5617

First published in 2006

Greenwood Press, 88 Post Road West, Westport, CT 06881
An imprint of Greenwood Publishing Group, Inc.
www.greenwood.com

Printed in the United States of America

The paper used in this book complies with the
Permanent Paper Standard issued by the National
Information Standards Organization (Z39.48–1984).

10 9 8 7 6 5 4 3 2 1

For Jean
For the many years of love, companionship, and shared
concern for animals, from London Zoo to Vancouver Island

Contents

Preface

There are many examples of the amazing evolutionary adaptations in the animal kingdom that aid survival, but the ability of animals to hibernate is one of the most incredible. Most vertebrates must keep warm, because hypothermia or loss of body heat is potentially fatal, but some have evolved to withstand a massive drop in temperature, and even to survive freezing. To avoid inhospitable conditions they enter a state of suspended animation known as hibernation. Cold to the touch, their hearts beat infrequently and they may stop breathing for long periods. They appear lifeless, but can survive in this condition for several months, some even for most of the year.

The long winter sleep of the northern hibernators is triggered by lowering temperatures, shorter days, or reduced food supplies, and in some species by an "internal clock" that controls their activities. In contrast, some animals sleep to avoid the hot and dry conditions of midsummer, which is called estivation. Their lowered metabolism reduces energy consumption, but even deep sleep must be fuelled to sustain life, and the hibernators store fat internally or have food caches nearby. As freezing is fatal to hibernating mammals, they awaken and shiver to warm up a few degrees when their temperature drops too low, but the cold-blooded fish, amphibians, and reptiles may freeze and die when exposed to frost, so several have developed unique survival skills. These include the production of anti-freeze compounds to protect their cell water, and the ability to lower the temperature at which they begin to freeze.

This is an account of the animals that are able to survive when their body temperature drops well below the normal danger level. It redefines hibernation, and includes animals such as the bears, which have not traditionally been considered hibernators.

Introduction

A major achievement in the evolution of animal behavior and physiology is the ability of certain animals to avoid seasonally adverse conditions. In much of the world there are times of year when their environment is too inhospitable for many species, so they must protect themselves from the elements. If migration to a better place is impossible, as it certainly is for the amphibians, reptiles, ground squirrels, and bears, they must take other precautions, and when it is too hot or too cold, or their food supplies dwindle, they become inactive and insensible in varying degrees until conditions improve. They are said to be torpid, and their periods of torpidity or deep sleep may last several weeks or months; in extreme cases up to three-quarters of each year. Torpidity is synonymous with hibernation and when long-term torpidity occurs in winter it is indeed called hibernation, while the term "estivation" is used specifically for summer torpidity. Hibernation is triggered by lowered temperatures, shortening days, and shortage of food; estivation by increasing temperatures and aridity.

When they could sleep in a relatively safe place to avoid the winter or summer extremes, nonmigratory animals were able to colonize inhospitable zones that were otherwise uninhabitable year-round. They could miss unpleasant conditions in relative safety and then continue their normal activities during more favorable times when food was plentiful. Consequently, the hibernators live in temperate latitudes, in the world's deserts and at high elevations even in the warmer regions of the world. Estivating species can be found in temperate zones also, and in the tropics where they become dormant to avoid the hot and arid conditions of midsummer. However, unless it is genetically disposed to dormancy, an animal cannot simply become torpid at will to escape unpleasant conditions. If it is not "programmed" to avoid bad weather and food shortages, and cannot as a result maintain its core temperature or retain its body water, death is inevitable. Hibernation also provided temperate-zone animals with the opportunity to raise their vulnerable babies in a den during the winter, safe from predators and the extreme climate and ready to take advantage of the arrival of spring.

Deeply torpid animals are in a state of suspended animation and may appear dead. They are usually limp and helpless, are cold to the touch, and their senses are very diminished. Their endocrine glands shrink, their metabolism is significantly reduced, and their respiration and heart beat are barely discernable. The body temperature of most species drops close to that of the environment, almost to the point of freezing, and in a few cases actually below it, when special adaptations prevent them from freezing solid, which is normally considered fatal to vertebrates.[1] When they are dormant, the endotherms or warm-blooded mammals and birds resemble the ectotherms (the reptiles and amphibians), but although their body temperature drops significantly they are still in control of their metabolism, unlike the ectotherms, which cannot produce their own warmth. Consequently, when a torpid mammal's temperature reaches a certain point close to freezing, called the set point, it generates heat to increase it a few degrees. As they cannot produce their own heat the ectotherms risk freezing and death, except for a few that have also evolved other strategies to survive being partially frozen. Their lowered metabolism requires less energy, which is provided by the food or fat they have stored in preparation for their long sleep. In addition to the long-term hibernators, many animals also practice short-term or daily torpidity, becoming lethargic for part of each day when they are resting, and in this way they conserve energy.

Although the cold-blooded reptiles and amphibians sleep solidly when the temperature is low, where the mammals are concerned, hibernation is not uninterrupted sleep. It may last for several days or weeks, but is broken by short periods of activity when the animal's body temperature returns to normal and it may even venture outside its den or hibernacula. Nor is long sleep necessarily lost time, for several studies have shown that compared to similar nonhibernating species the animals that sleep for half of each year live longer, and have the benefit of being active only during favorable times.

Hibernation is closely associated with energy production and its utilization. The evolution of torpidity hinged upon the amount of energy needed to maintain a mammal's low temperature all winter versus the savings less the energy costs of the warm-up. Short periods of torpor involving a large drop in temperature are uneconomical, due to the energy required to regain normal temperature. Most hibernators are therefore small, because they can better withstand the frequent cooling and then rewarming that is associated with torpidity, and which is believed to account for most of the energy used during the long sleep. Small body size provides greater surface area per volume and therefore greater opportunities for heat loss if animals remained active, but less energy is needed for their sleep and for the warm-up. Smaller animals have the most effective energy reduction during hibernation, illustrated by the fact that the metabolic rate of a hibernating marmot weighing 10 pounds (4.5 kg) is 5 percent of its normal active rate, but is only 1 percent in the ⅙-ounce (4.5 g) pipistrelle bat.

It is uneconomical for large-bodied animals to sleep deeply and undergo a major reduction in their breathing, heart rate, and temperature, due to the large amount of energy needed to rewarm their bodies afterward, and bears are able to sleep for many months only because their temperature does not drop significantly. The largest mammal to hibernate with a body temperature close to freezing is the hoary marmot, which weighs up to 18 pounds (8 kg) when it has stored fat for the winter. Small animals like the shrews and hummingbirds are also the main proponents of daily torpor. They need lots of food to power their rapid metabolic rate and high body temperature; and to conserve energy when they are resting, when food is short or the temperature is low, they enter a state of torpidity for part of each day. The northern insectivorous bats, which hibernate for the winter, also become torpid when they are resting during the day in summer, whatever the temperature.

The physiological adaptations of hibernating animals have numerous potential applications for human medicine, and many aspects are currently being actively researched. Hibernating painted turtles and pond sliders can survive without breathing for several months, whereas oxygen deprivation in humans may cause brain damage after just four minutes, and death soon afterward. The anti-freeze compounds that the partially freeze-tolerant species produce to protect their intracellular fluids while allowing the fluids between the cells to freeze have great potential for extending the life of human organs destined for transplant surgery. Hypothermia—abnormally low core temperature—causes respiratory failure, cardiac arrest, and death in humans and nonhibernating mammals, yet hibernating animals can survive in their comatose state when their temperature drops close to freezing, and in some cases even below. This has potential value for a number of situations, such as stabilizing soldiers injured on the battlefield and the victims of traffic accidents during their transportation to hospital. The fact that many hibernating mammals can maintain their normal blood sugar levels on a diet comprised solely of stored fat is also of great interest to human medicine. The blood sugars of active Arctic ground squirrels decreased when they were starved for six weeks in midsummer, whereas those of hibernating animals were maintained at a constant level. Wood frogs can tolerate extremely high blood sugar levels, but do not suffer the consequences that afflict human diabetics when their levels rise.

Bears appear to offer more advantages for human medicine than any other animals. Mammals, including humans, need physical activity to maintain normal physiological functions, and immobilization results in dysfunction of the organs and tissues, and severe bone and muscle degeneration. However, bears that spend the winter in a cramped den, and do not eat, drink, defecate, or urinate, do not show these adverse effects. Also, their cholesterol level is double its summer rate, as a result of metabolizing body fat for energy, yet they do not suffer the usual human afflictions such as hardening of the arteries and cholesterol gall stone formation, which result from high cholesterol levels. The knowledge that bears can maintain their body water balance without drinking for several months, and reduce the urine entering their kidneys by 95 percent while sleeping, is of great interest to researchers studying human kidney disease. Sleeping bears can also maintain their muscle tissue even on their diet of stored fat because they recycle their metabolic wastes, and the nitrogen produced when urea is broken down is used to build protein for maintaining their muscles and organ tissues.

Polar Bears *A female polar bear and her cubs on the Arctic ice shortly after leaving the den in a snowbank where the cubs were born 3 months earlier. Female polar bears den purely to protect the cubs from the elements, while the males remain out on the ice all winter hunting.*
Photo: Courtesy USFWS

Although the very purpose of hibernation is to provide a suitable environment in which animals can avoid bad weather and the lack of food, it is in fact a dangerous time and many do not survive. Inactive and unable to move quickly from a threat, they are at the mercy of the weather and predators, and must find a safe place where they are protected from both. Snow is the ally of the northern and high-elevation species, providing an insulating blanket between their hibernacula and the frigid weather. The lack of snow and the resultant frost penetration are particularly harmful to the terrestrial amphibians, which sleep under a thin covering of leaves and soil; and aquatic amphibians, which spend the winter on the bottom of a pond, are vulnerable if all their water freezes during hard winters. Survival also hinges upon having sufficient energy, either in the form of internal fat reserves or external food stores, which may not last if spring arrives late. Natural disasters such as a rapid early thaw have drowned ground squirrels in their dens, and on the Barren Grounds gaunt and hungry grizzly bears emerge from hibernation before the ground squirrels and dig them out of their burrows. In studies that have counted the number of ground squirrels in a colony at the start of their hibernation, the total that emerged was always much lower.

There are several inaccuracies and inconsistencies associated with hibernation. It is said, for instance, to be comparable to prolonged fasting, but this is certainly not true of the many species of external food storers, which arouse regularly to eat. The internal storers have in fact also prepared for their long sleep by laying down fat which they then use for their energy requirements, so their "fast" is not the same as abstinence from eating for weight loss purposes or religious reasons. Neither group of animals goes without sustenance during its long sleep. Hibernation in mammals has also been likened to hypothermia, an unplanned and unregulated condition in which the body fails to maintain adequate production of heat under conditions of extreme cold and lack of protection, resulting in a low core body temperature that is fatal at a certain point. Many mammals have indeed evolved the ability to lower their core temperature to just above freezing without harm, but with a difference. It is regulated and under the animal's control, and both hibernation and daily torpor in mammals could therefore be termed regulated hypothermia.[2] However, torpor in the cold-blooded reptiles and amphibians is perhaps more typical of hypothermia as it is unregulated, although it is certainly not unplanned, but the result of millions of years of evolution.

The most inconsistent aspect, however, involves the very definition of a hibernator. The behavior and physiology of the many animals that practice hibernation differ considerably according to the species, and their torpidity can take several forms. There is the deep sleep of the rodents and bats in which the body temperature drops close to freezing, and they awaken slowly. In contrast the carnivores sleep lightly and can arouse quickly to defend themselves and their cubs. There are significant differences between the hibernation processes of the amphibians and reptiles, which are totally dependent upon the environment for their warmth, and the mammals and birds, which generate their own heat. Then there are the animals that store fat internally for their energy, and those that stock a larder with food. Currently, which of these animals should be called hibernators depends entirely upon an outdated definition.

According to present philosophy, "true" hibernators are animals like the ground squirrels and hamsters, which experience a significant drop in temperature and are consequently chilled and comatose, and warm up and awaken slowly. Therefore, bears and other carnivores do not qualify, as their body temperature does not drop significantly and they are light sleepers that awaken quickly. Consequently, their long sleep has confusingly been called torpidity, dormancy, partial hibernation, winter sleep, and other names, rather than hibernation. Bears sleep for up to seven months without eating, drinking, defecating, or urinating, and it has been said that they are not really hibernating but are just in a state of torpor. They are unquestionably torpid, and torpidity is synonymous with hibernation. Their drop in core temperature, although small compared to the ground squirrels, does reach 88°F (31°C) from their normal 100°F (37.8°C), which is well below the survival level of most nonhibernating species,[3] and which is therefore hypothermia. In humans a core temperature drop to 95°F (35°C) is considered dangerously low and potentially fatal.

Bears are in fact very efficient hibernators, perhaps even more so than the ground squirrels and hamsters, which awaken regularly to raise their body temperature, to urinate, and to feed from their stores. Bears have a higher temperature for two reasons. Pregnancy, birth, and nursing all occur in the den during hibernation, and this could not be achieved if they were on the brink of freezing and their body processes were barely functioning. Second, due to their large size, the high cost of energy needed to rewarm their bodies from a much lower temperature would be uneconomical. Their sleep is just one of the many forms of dormancy, differing from the long winter sleep of other mammals—especially the disturbed sleep of those that awaken to feed—and the uncontrolled sleep of the ectotherms; but it is undoubtedly hibernation. Bears obviously do what is best for them—for their big bodies and their unique breeding behavior—and they do it very successfully. Similarly, a raccoon that normally weighs 22 pounds (10 kg) and increases this by half with fat stored for its winter sleep (which can last for several months in the north) is obviously a hibernator, and this applies equally to other long- but light-sleeping carnivores such as the badger and raccoon dog.

For too long the animals practicing long-term torpor have been categorized under an outdated and unsuitable definition of the term "hibernation." With the increased knowledge of torpidity and its mechanisms it is unacceptable to continue excluding obvious hibernators like the bears, the badgers, and the skunks simply because they do not fit the existing definition. After all, the very word "hibernation" is derived from the Latin *hibernatus* meaning "to pass the winter." Consequently, for the purpose of this book, hibernation is defined as "long-term torpidity to escape unfavorable winter conditions and to conserve energy, while surviving upon body fat or external food stores."[4] Estivation is "long-term summer sleep to avoid hot and dry weather and to conserve water and energy, while surviving upon body fat." Daily torpidity is considered an appropriate name for semilethargic, energy-saving sleep of a few hours' duration.

Notes

1. Researchers recently claimed to have frozen laboratory rats down to 19.4°F (−7°C), and then revived them without ill effects.

2. Controlled hypothermia has been used to temporarily lower metabolism during human surgery.

3. Several nonhibernating mammals, such as the sloth and platypus (plus the hibernating echidna), have very low normal body temperatures, similar to the bear's torpid level.

4. The term "hibernation" is also used by cardiologists to describe a condition in which segments of the heart muscle appear to be dead, but can be revived.

1 Suspended Animation

It is uncertain how hibernation evolved, but it seems likely that an animal's ability to become torpid at will is not a newly acquired characteristic. Rather, it is a primitive trait that has been lost by animals residing in warmer climates. However, to spend time in a state of suspended animation is now a very necessary aspect of the lives of many animals. It not only protects them from the elements, saves energy, and allows them to reside year-round in regions where they could not otherwise survive, but has evolved to the stage where a period of torpidity is essential to induce breeding activity and ensure reproductive success, because the processes of ovulation and spermatogenesis are inhibited when animals cannot hibernate.

The various levels or types of torpor practiced by animals range from the supreme conditions of deep suspended animation to the simple lowering of the body temperature a few degrees. In addition, there is the form of short-term regulated dormancy that is known as daily torpor, which lowers energy consumption. Animals that have evolved to hibernate can survive the lowering of their core temperature to a level that would kill nonhibernators, although even they cannot survive if their whole body is frozen solid.

■ THE TYPES OF TORPIDITY

Hibernation

Hibernation is long-term torpidity or dormancy to escape unfavorable winter conditions and to conserve energy, while surviving upon body fat or external food stores. It is always a seasonal activity and its start is as predictable as the arrival of winter, with animals prompted to commence their sleep by certain triggers that include their internal clocks, shorter days, lowered temperatures, or lack of food. These affect their endocrine glands, influencing the secretion of hormones that initiate the process. Both endotherms (warm-blooded animals) and ectotherms (cold-blooded animals) hibernate.

In the endotherms it is an innate response to certain conditions, but actually commences in anticipation of the event—the impending arrival of unfavorable times. It is characterized by long periods of sleep with lowered body temperature and greatly reduced metabolic rate. In the mammals and birds it is a regulated activity, and they are always in control of their body functions,[1] their metabolism being lowered to a level that conserves energy and normally allows their fat or food stores to last for the duration of their dormancy. The metabolic rate during deep sleep is reduced to about 5 percent of the basal metabolic rate (BMR)—the rate of nonhibernating animals at rest that is needed to support basic functions such as breathing and blood circulation. Even including the regular awakening from torpor and rewarming the body, which uses most of a mammal's stored fat, the metabolic rate during hibernation averages only about fifteen percent of the BMR.

Mammals are not torpid for the whole period of their hibernation, but arouse periodically, and their metabolism then increases until they once again have a normal body temperature and are said to be normothermic. In some species these arousals follow a regular pattern and must therefore be controlled by internal body rhythms. Hibernators also arouse if their temperature reaches their "set-point," which gives warning of impending freezing and at which time they shiver to raise their temperature.

Mammalian hibernation takes several forms. Many rodents and bats sleep deeply with their temperature close to freezing, whereas the carnivore hibernators are light sleepers that can arouse quickly. The echidnas, which have a much lower body temperature and therefore BMR than most mammals, are almost reptilian in their inability to keep warm in cold weather. There are species that store fat internally for their energy, and those that stock a larder with food. The female polar bear hibernates, but the male does not, yet in the other northern species of bears both sexes hibernate.

Hibernating mammals range in size from bats to bears, but there is a definite size limit to the species that hibernate at very low temperatures, the maximum being 18 pounds (8 kg), reached in the hoary marmot when it has fattened up for its long sleep. For species that hibernate successfully at higher body temperatures there is no weight limit, as the largest northern land mammal—the male Kodiak bear—has evolved this ability. In fact, the largest mammalian hibernators are carnivores, and include the four northern bears (the polar bear, brown bear, American black bear, and Eurasian black bear), plus smaller species such as the raccoon, both kinds of northern badger, and the raccoon dog. However, most hibernating mammals actually weigh less than 2 pounds (900 g), as smaller species benefit more from dormancy during cold weather. This is because the surface area-to-volume ratio of an animal's body increases as size decreases, allowing greater loss of heat, so they must produce extra heat to compensate when it is cold, but the extra food needed to "power" this heat is usually unavailable in winter and the animal is forced to take evasive action.

Bats are the smallest hibernating mammals, with the eastern pipistrelle (*Pipistrellus subflavus*) and the common pipistrelle[2] (*P. pipistrellus*) barely reaching ⅙ ounce (4.8 g) in weight, and several other *Vespertilionid* species weighing only ⅕ ounce (5.6 g). The southern birch mouse (*Sicista subtilis*) also weighs ⅕ ounce and is therefore the smallest terrestrial (nonflying) mammal known to hibernate. Next in size are the marsupial little pygmy possum (*Cercartetus lepidus*), and several pocket mice (*Perognathus*) which weigh just ¼ ounce (7 g). Male honey possums (*Tarsipes rostratus*) also weigh only ¼ ounce and have been found huddled torpid in their nest during cold and wet weather, but it is unknown if this was just daily torpidity or part of longer-term hibernation. Such uncertainty also applies to another small marsupial, the feather-tailed glider (*Acrobates pygmaeus*), which weighs ⅜ ounce (12 g). The smallest primate to sleep for long periods is the pygmy mouse lemur (*Microcebus myoxinus*), which weighs just over 1 ounce (28.3 g).

Although many birds, including some quite large species, practice daily torpidity, only the poorwill (*Phalaenoptilus nuttalli*) is known to hibernate, but it is just a featherweight, weighing only 1½ ounces (42.5 g).

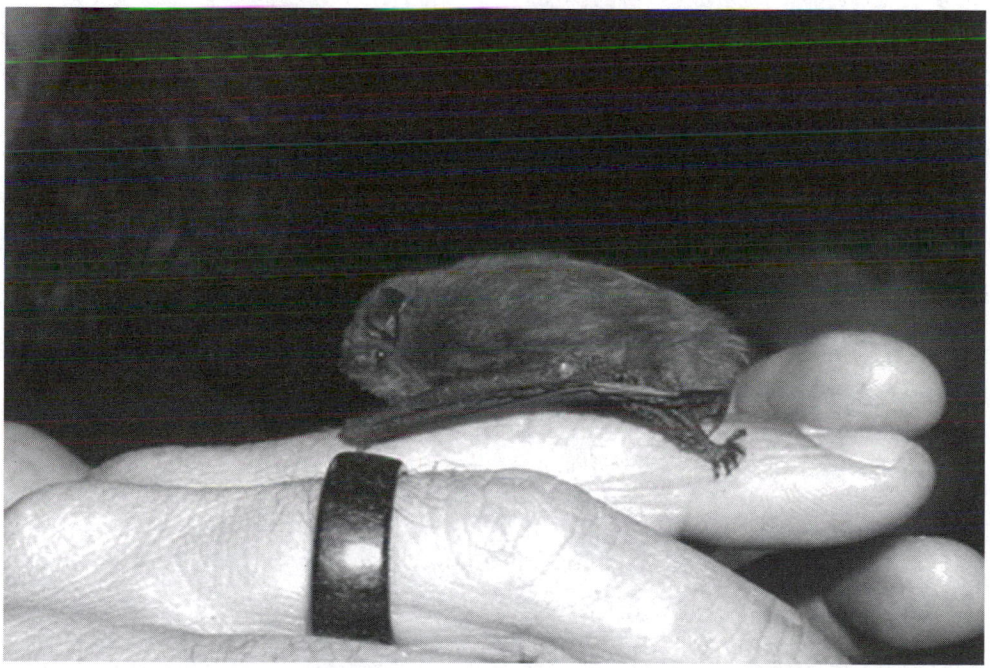

Common Pipistrelle *This tiny bat is the smallest hibernating mammal, averaging 1½"* *(4 cm) in length and just ⅙ oz (4.8 g) in weight. In addition to hibernating from early* *November to early April in northern Europe, it also becomes torpid daily when roosting* *to save energy.*
Photo: Markus Nolf

The major differences between hibernators are actually not those between the light-sleeping bears and the deep-sleeping rodents, but between the endotherms and the ectotherms. Mammals are mostly forced into hibernation by the lack of food, not low temperatures, whereas ectotherms must hibernate or die when it gets cold. Dormancy in the warm-blooded animals is controlled sleep, whereas that of the cold-blooded species is uncontrolled and enforced. Unable to produce their own warmth to counter the cold, amphibians and reptiles are totally at the mercy of lowering environmental temperatures, which chill them to sleep. They cannot produce heat to awaken on their own accord in spring, and must rely on the warming of the environment and basking in sunlight, or lying on sun-warmed rocks, to raise their metabolic rate. They also differ from many hibernating mammals in the manner in which they store their winter energy supplies, as this occurs only within their bodies. None store food externally for the winter, as they cannot awaken to eat their stored foods, and in fact they do not arouse at all during their long sleep unless it gets unseasonably warm. When this happens, however, it only affects the surface-hibernating species such as the wood frog (*Rana sylvatica*) and spring peeper (*Pseudachris crucifer*), and not the deep-burrowing spadefoot toads (*Scaphiophus*) or the aquatic leopard frog (*Rana pipiens*) on the pond bottom beneath a thick layer of ice. One advantage that several ectotherms have over the mammals is their ability to produce anti-freeze which allows them to survive being partially frozen for short periods.[3] The northern ectotherms, which are forced into hibernation by lowered temperatures, range in size from many tiny amphibians—frogs, toads, and salamanders—to the giant alligator snapping turtle, which has a record weight of 219 pounds (99 kg). They include the lungfish and carp, spadefoot toads and spring peepers, garter snakes and hog snakes, Gila monsters and box turtles.

Several subcategories of hibernation are recognized. Obligate or spontaneous hibernators are those that enter torpor regardless of the amount of food and water available, being driven by their circannual rhythms—their internal biological clocks. Facultative hibernators are those in which the starting time is optional, and they stay out if the winter is mild and food is still available. Predictive hibernators are those that begin their sleep in response to shorter day length, anticipating the approach of winter.

The term "hibernation," is normally used for winter dormancy, when there is a marked seasonal temperature change, and "estivation" refers to summer dormancy when the shortage of food and water are the main reasons for a long sleep. Many animals, especially the moist-skinned amphibians, must avoid dry conditions, and summer estivation is therefore necessary in some regions to prevent their desiccation. When their habitat experiences inhospitable extremes in both summer and winter—such as the deserts of the southwestern United States—some animals even estivate and hibernate in the same year, the two periods of torpidity running consecutively.

Estivation

Estivation is long-term summer torpidity or dormancy to avoid hot and dry conditions while surviving upon internal stores of water and fat. It involves an abnormal lowering of the metabolic rate, similar to that experienced during hibernation, so that animals can survive inhospitable conditions, especially lack of water. All their systems slow down, and estivation is virtually indistinguishable from hibernation, comparable in purpose and attributes although differing in some of its mechanisms, including anti-dehydration measures such as water storage and cocooning in a waterproof membrane to prevent water loss. However, estivating animals generally have a higher body temperature and metabolism than the winter hibernators as the ambient temperature is higher. They survive on their stored fat and do not store food externally, except for possibly some small desert species such as the kangaroo rats and jerboas.

Estivation occurs mainly in regions experiencing very hot summers[4] and low rainfall, and the subsequent drying of the soil, ponds, and watercourses. This can happen in midsummer in temperate regions such as the Canadian prairies and the arid lands of the southwestern United States, or in the seasonally dry grasslands and deserts of tropical latitudes. It occurs in both warm-blooded and cold-blooded animals, although it is most common in the latter, especially the amphibians, which must protect their moist skins from desiccation. In addition to burrowing deep to avoid the hot surface conditions, they store water and create waterproof cocoons to retain their moisture. Estivating amphibians include the African bull-frog (*Pyxicephalus adspersus*) of the sub-Saharan grasslands, and South Africa's rain frog (*Breviceps gibbosus*). In North America the greater siren (*Siren lacertina*) and the spadefoot toads (*Pelobatidae*) of the plains and the Great Basin estivate. Among the reptiles it is also the desert species that estivate to avoid intense heat. The spurred tortoise (*Geochelone sulcata*), the desert tortoise (*Gopherus agassizi*), and the desert box turtle (*Terrapene ornata luteola*) burrow to escape the extreme heat of summer and then low temperatures in winter, and the latter may spend three months estivating very soon after its five-month-long winter hibernation.

Among the mammals, the desert hedgehogs of the Sahara, the Gobi, and the Great Indian Desert all estivate in midsummer. Many desert rodents, including the white-tailed prairie dog (*Cynomys leucurus*), the yellow pine chipmunk (*Eutamias amoenus*), the Uinta ground squirrel (*Spermophilus armatus*), and the desert jerboa (*Jaculus jaculus*), estivate to avoid high temperatures. Several small marsupials of the arid center of Australia, including the narrow-footed marsupial mouse and the fat-tailed dunnart, sleep during the torrid midsummer heat. In contrast, the tenrecs and dwarf lemurs estivate in Madagascar's winter, which is the island's dry and slightly cooler season.

Although it is also basically a seasonal activity, estivation is more erratic than winter hibernation, due to the uncertainty and irregularity of drought, but if estivators can stay moist they may remain torpid for very much longer than hibernators, which use up their energy stores at a faster rate during low temperatures and especially when rewarming their cold bodies. The African lungfish may sleep for three years, and spadefoot toads in the American Southwest normally spend eleven months in torpidity, but when conditions are unfavorable, due to a long drought, they have estivated and hibernated continually for two years.

Like the winter hibernators, the ectotherms, which sleep to avoid arid conditions, must have sufficient water and energy supplies to maintain their low metabolic rates. They must also protect themselves from the elements, avoiding dehydration when they are surrounded by hot and dry sand or soil, or baked in the mud at the bottom of a dried-out pool, so water retention is critical for them all. They need a site that provides safety from predators and protects them from the elements and water loss, and they have evolved specialized behavioral and physiological adaptations for survival. Frogs estivate in a crouched position with their limbs tucked beneath them, to reduce their body surface area and cut down on the loss of water by evaporation. Their physiological adaptations include water storage and greatly reduced breathing to counter the loss of fluids, and they may even stop breathing for short periods. Frogs and toads enter estivation with an extra supply of water stored in their bladders, and in some species evaporative water loss through their permeable skins is countered by making a waterproof cocoon with layers of their shed skins; the lungfish secretes mucus that dries into a protective layer around it. But frogs and toads are very tough animals; they can survive the loss of up to 60 percent of their total body water during estivation, and although they emerge looking very shrivelled they can absorb water quickly and soon return to their normal shape. Estivating ectotherms must arouse immediately when the rains begin, for they have a very short season to complete their breeding cycle and prepare for hibernation. They must mate and lay their eggs quickly, allowing sufficient time for tadpole growth and metamorphosis, and must feed voraciously to store sufficient reserves for their next session underground.

Daily Torpor

Daily torpor is energy-saving, semilethargic sleep, usually of about half a day's duration only. It is interspersed with periods of activity and normally occurs at a definite time of day and season. It is a less profound state of dormancy than full hibernation; a resting period caused by self-induced reduction of body temperature and metabolic rate. In temperate climates it is an animal's response to periods of food shortage or low temperatures by lowering its metabolism and thereby reducing its demands for energy, and can be entered without preparation simply by reducing its body temperature. Surprisingly, it also occurs when it is warm and there is no shortage of food in species with high metabolism and large food requirements such as the shrews and hummingbirds, where it is also concerned with saving energy. Daily torpor is incorporated into the normal circadian rhythms of sleep and activity, but may not be restricted to an animal's usual resting period.

Daily torpor, which has also been called shallow or rest-phase torpor, is more widespread than previously suspected. It can occur at any time of day and any time of year, and like hibernation is mainly a strategy of animals with a very active lifestyle, small body size, and high metabolism. They conserve energy by lowering their temperature every night, or day in the case of nocturnal species, irrespective of the air temperature, and even in the tropics.

In mammals it has been reported in bats, shrews, skunks, badgers, deer mice, gerbils, tenrecs, sugar gliders, marsupial mice, honey possums, dasyures, and dwarf lemurs. The sloths, weighing about 20 pounds (9 kg), are believed to be the largest mammals to undergo daily torpor, and it is also suspected in the duck-billed platypus and echidna. The bumblebee bat (*Craseonycteris thongyonglai*), weighing just 1/14 ounce (2 g), is a tropical species that may become torpid daily like so many small insectivorous bats, in which case it would share with the Etruscan shrew (*Suncus etruscus*) the title of the smallest mammal to undergo daily torpidity, and there are no smaller mammals in the world than these.

Birds also practice energy-saving daily torpor, their body temperature dropping several degrees when they are roosting. It occurs in many small birds including pigeons, hummingbirds, swifts, martins, swallows, chickadees, manakins, sunbirds, todies, poorwills, white eyes, and honey eaters. Several larger species also regularly become torpid when resting, including the roadrunner, the black vulture and turkey vulture, and the red-tailed hawk, and it is assumed to happen in other species. Nestling birds become torpid to conserve energy when the weather is cool and their parents have difficulty finding enough food for them. Birds that undergo this daily chilling generally arouse the following day or night to eat, although nestling swifts have been left comatose for several days in cold weather when insects were scarce. The basal metabolic rate is related to mass, with small birds expending more energy per unit of weight than large ones.

Hummingbirds and chickadees, with high metabolic rates and therefore high energy use, cannot survive without food for very long, and must feed soon after arousing to replenish their energy and be prepared for the next torpid period,

Black-capped Chickadee *In the northern forests in mid-winter the chickadee feeds for barely 8 hours and then roosts for 16 hours in sub-freezing temperatures. To make its small reserves of fat last all night it becomes temporarily hypothermic, its body temperature dropping 18°F (10°C) from its normal 107.6°F (42°C).*
Photo: Bruce MacQueen, Dreamstime.com

because unlike hibernation, short-term torpidity does not involve using extra internal food supplies stored for that purpose. The proponents of daily torpor rely on feeding during their active periods to have sufficient reserves for their sleep, when they use less energy due to their lowered metabolism. For those that do not experience a large drop in temperature during their daily torpidity, the other physiological changes characteristic of hibernation (lowered respiration, heart rate, and blood flow) are also minimal, but in species such as the insectivorous bats, which have a major temperature reduction when resting by day, the only difference between daily torpor and hibernation is its duration.

A constraint for daily torpor is the cost of energy required to return the body to its normal temperature. Basking in the sun, or solar heat gain, is used by many animals (including the frogmouth, elephant shrew, fat-tailed antechinus, and the common poorwill) to achieve warm-up at less cost to their own energy reserves. But even without the benefit of solar heat the savings in energy resulting from their semilethargic sleep more than compensates for that expended for the warm-up. Despite their very low temperatures during torpor, bats can recover their normal functions quickly through shivering and shaking their wings. Daily torpor is not generally practiced by pregnant or lactating mammals or incubating birds, due to the reduced escape potential and lowered milk production in mammals and lowered heat transfer to their eggs by birds.

■ PREPARING FOR HIBERNATION

Mammals prepare for hibernation well in advance of the actual event, with the accumulation of fat reserves or food stores and the preparation of the den, which must provide protection from the elements and from predators. This den or hibernacula may be a burrow with a sleeping compartment dug just for the purpose by a ground squirrel, or a burrow system lived in throughout the year, and often for many generations, by a family of badgers. It may be a nest made in a tree cavity by a dwarf lemur, a hole dug beneath the roots of a fallen tree by a black bear, or the female polar bear's tunnel in a snowbank.

Generally, the purpose of the hibernacula is to provide a winter home that protects them from frost. For the ground squirrels deep in their burrows the nest temperature usually remains a few degrees above freezing.[5] The polar bear's snow den may remain at the freezing mark, whereas hibernating black bears may sleep in a den which is open to the winter elements. Bats that hibernate in caves generally prefer higher temperatures, between 37°F (2.8°C) and 42°F (6.1°C) with high humidity.

Bears hibernate individually, and the kangaroo rats, woodchucks and dormice are also loners, but many mammals are very colonial and huddle with several members of their own family or join many others. Hoary and alpine marmots sleep in groups of up to fifteen animals, and skunks and raccoons huddle together in family groups to keep warm and reduce their demand for energy. The insectivorous bats are the most colonial hibernators, some species congregating in the hundreds of thousands in traditional caves that are used for many years.

With few exceptions the ectotherms hibernate just where they happen to be when they are prompted by the weather to take action. Spadefoot toads burrow deep into the sandy soil, box tortoises and salamanders push beneath piles of leaf litter and loose soil; and fish, aquatic frogs, and turtles rest on the bottom of their pond. Exceptions include burrowers such as the gopher tortoise, which digs its own burrow and may use it for several winters, and also for summer estivation; and the garter snakes, which use the same limestone sink-hole for generations. This could be several miles from the sloughs and lakesides where they have spent the summer.

To sustain life for several months, however inactive they may be, the hibernators must have sufficient reserves, both for their period of lethargy (and in the endotherms for their periodic awakenings) and for the reproductive period that follows their emergence. Some northern amphibians breed in icy cold water, long before food is available again, and ground squirrels and marmots often emerge while there is still snow on the ground. So in preparation for their seasonal sleep there is a period of special physiological adjustment (the storing of food reserves within their bodies) or of behavioral adjustment (the gathering and storing of food in their larders).

Animals generally do not begin to hibernate unless they have made the necessary arrangements. They must have sufficient supplies to last until the return of warmer weather, and dormancy can be suspended until they are fully prepared. Black bears do not enter hibernation when they have not laid down sufficient internal fat reserves, which places them in a difficult situation as they cannot possibly survive the winter if they do not den. Some mammals will not begin their sleep until they have found or prepared a suitable hibernacula, like the captive hedgehogs and dormice that remained active in the cold until nest boxes and nesting material were provided for them. Warm-blooded hibernators either store food internally in the form of white or brown fat, or externally in "larders" stocked with foods which they eat when they awaken periodically. Some species (ground squirrels and dormice) which gain considerable weight with fat for the winter, make doubly sure they have enough so they may also have a store of food for when they awaken.

Wherever their food is stored, it must be sufficient to last not only for the duration of hibernation but also for the awakening and warming-up process and the immediate breeding season, which usually begins right away, at a time when food may not yet be available.

The internal fat storers include the northern bats, some dormice, flying squirrels, marmots, ground squirrels, prairie dogs, bears, dwarf lemurs, pygmy possums, hedgehogs, jumping mice, birch mice, the poorwill, and all the ectotherms or cold-blooded vertebrates. Ground squirrels and bears build up a layer of subcutaneous fat around their bodies, with an extra thick layer on the rump. A female polar bear must gain 440 pounds (200 kg) of fat prior to commencing hibernation, whereas the tiny mountain pygmy possum survives the winter on ½ ounce (14 g) of stored fat. Several species store fat in their tails, like the lesser mouse lemur, which weighs only 1½ ounces (42 g), but its tail swells to 1½ inches (4 cm) in diameter. Storing adequate fat is usually dependent upon good eating in the few weeks prior

Chipmunk *Hibernating mammals either store fat internally to power their long winter sleep or they fill larders with seeds, grains, nuts, and berries and awaken regularly to feed. The chipmunk is a food storer, and to aid its food gathering in the fall it has large cheek pouches to carry food back to its underground storage chamber.*
Photo: WizData.Inc., Shutterstock.com

to their long sleep, usually a time of plenty in the temperate regions with rich foods available, such as nuts, grains, and berries for the dormice, and salmon for the grizzly bears. The tail is also a favorite site for fat storage in the cold-blooded hibernators. The Gila monster survives for several months of estivation and hibernation on its stored fat, most of which is in its thick, stubby tail; and several geckos also store fat in their tails. Snakes store fat in the body cavity called the coelom, which is between the body wall and the digestive tract, and wood frogs store fat and glycogen, which is converted to glucose when required as a cryoprotectant. Their stored fat is mainly for energy during the breeding season when they emerge, as it is too cold for their invertebrate prey to be available.

External food storage is only possible for small animals with an omnivorous or herbivorous diet that can store nonperishable items, and no purely insectivorous or carnivorous mammal stores a supply of animal protein for the winter. A polar bear needing a very large amount of meat or fat daily cannot store its winter supply, nor could an insectivorous bat, which is accustomed to catching hundreds of insects nightly. The external food storers include the hamsters, chipmunks, some ground squirrels, kangaroo mice and rats, pocket mice, and some dormice. They store grass, nuts, grains, seeds, tubers, and bulbs, foods that have dried naturally such as berries and possibly fungi; and the large common hamster may store cultivated root vegetables such as potatoes and carrots, which will keep in the coolness of their dens. To aid their food gathering some storers have cheek pouches into which they can stuff large amounts of food to carry back to their underground chambers, and they store an amazing amount of food. The larder of one tiny kangaroo rat contained 12 pounds (5.5 kg) of assorted seeds, and the common hamster's hoard of winter foods has weighed up to 200 pounds (90 kg), 100 times its own body weight.

■ TRIGGERING HIBERNATION

The commencement of hibernation is usually associated with the arrival of cold weather. This certainly applies to the ectotherms and to some mammals, but the cues that trigger their hibernation vary according to the species. Some are considered opportunistic hibernators which may become torpid at any time of year when conditions are unfavorable—generally low temperatures (or high temperatures in summer) or reduced food supplies. The pygmy possums (*Cercartetus*), the serotine bats (*Eptesicus*), and the free-tailed bats (*Tadarida*) are such animals. Others are triggered by changes in day length (the photoperiod), the shortening days of fall initiating torpidity in the dormouse and the hamster. An animal's biological clock plays an important role in hibernation, with the circannual rhythm (an internal self-sustained annual cycle) triggering the onset of hibernation and then arousal in some species, such as the white-tailed prairie dog and the woodchuck. Estivation is triggered by increasing temperatures and dryness, coupled with the drying of ponds and vegetation, and food becoming scarce.

Whatever the actual trigger, it is an indisputable fact that hibernators generally begin their sleep when it is cold, and reappear when the temperature begins to warm up in spring, but the timing may vary due to a number of circumstances. Irrespective of the temperature, and even when it has remained high, when food was still available and the light maintained at summer levels, thirteen-lined ground squirrels in the laboratory began to accumulate fat and then get drowsy in early September, as they were governed by their circannual rhythm rather than the environment. Another internal clock—the circadian rhythm—which has a daily cycle, is believed to be involved in the regular awakenings of hibernating mammals. The time of commencement may also vary specifically according to an animal's cold hardiness. The golden hamster waits for several weeks in quite cold weather before entering hibernation, which it then starts with a major temperature reduction. Gold-mantled ground squirrels (see color insert) also seem reluctant to start and have appeared on sunny days in the early winter even when the air temperature was well below freezing.

Mammals vary in the manner in which they start their periods of sleep. Hibernation may begin with alternating short periods of torpor and arousal before the main event begins, but during these times their body temperature does not drop significantly. Others go straight into deep torpor, although for the whole period of mammalian hibernation, however long it lasts, the level of sleep varies and is believed to be deepest in the middle of the period. However, although the term "sleep" is used to describe the inactivity of the dormant animal, it is not sleep as we understand it, as the brain waves that normally occur during sleep are absent in hibernation. In fact, brain activity virtually ceases, and the animal's metabolism is almost at a standstill, so it is not rejuvenating its body in the same manner as sleep does, which includes such benefits as maintaining health and maximum alertness and benefiting neural functioning and the immune system. It has even been suggested that the frequent periods of arousal during hibernation may simply be opportunities for the animals to sleep and benefit accordingly. Lacking control of their body temperature, ectotherms' activity fluctuates according to the environment and they may become torpid overnight and then aroused by the sun for several days or even weeks before the main period of total torpidity begins.

■ THE PHYSIOLOGY OF HIBERNATION

The physiology[6] of hibernation includes various aspects such as glandular modification, thermoregulation, metabolic rate reduction, internal fat storage, and the conversion of fat into energy. The actual main event of hibernation—the condition of torpidity—results from changes in an animal's thermoregulatory system, which affects the mechanisms of heat production. This occurs when the hibernation triggers (cold, shorter day length, lack of food, internal clock, etc.) influence the secretion of hormones by the endocrine glands, which initiate the process of hibernation. For example, secretions of the pineal gland (which produces melatonin that is involved in sleep cycles) increase in low light and are believed to influence the onset of hibernation. Melatonin also causes the growth of thick winter coats during the shortening days of the fall for animals that do not hibernate, and the reduced production of melatonin in spring results in their shedding. The thyroid is concerned with the regulation of the metabolic rate, and its involution or shrinking and reduced secretions prior to hibernation depress activity. But it is the pituitary, the master gland of the endocrine system, that is most involved with the processes of hibernation, as it controls the activities of other glands. Located at the base of the brain, it is involved in the deposition of fat, the lowering of body temperature, and the slowing of the heartbeat and respiration. It triggers the release of insulin into the bloodstream, resulting in a lowering of the glucose needed by the animal, and its storage as glycogen in the liver and muscles for future use; this conversion of glucose into glycogen prepares the animal for hibernation. The hibernating animal's temperature is initially affected by this glandular activity. Heat production is reduced and the temperature drops, lowering the metabolism. As this happens the animal becomes less mobile and eventually enters a state of torpidity or dormancy. Consequently, the typical warm-blooded creature has become virtually heterothermic—allowing the environment to dictate its body temperature, to a point. Whereas both hibernation and estivation directly involve lowered temperatures and reduced metabolic rates, the mechanisms that actually cause the metabolic rate to slow and produce torpor vary. Species that undergo daily torpor rely on the reduction of their body temperature to reduce their metabolism, whereas longer-term hibernators rely more upon metabolic inhibition and then the lowering of body temperature to reduce their metabolism.

However, in the mammals (and one bird) hibernation is a controlled condition and they regulate their body temperature even when in deep torpor. They do not just shut down and lie sluggish at the mercy of the environment. Their temperature drops only to a prearranged level called the "set point," below which the animal arouses and shivers violently to raise its temperature—a safety mechanism to ensure it does not freeze, which is fatal to birds and most mammals. The set point differs among species and is related to body size and the environment. When the mammal warms up it must use self-generated heat, burning fat to provide this energy. Similarly, endotherms which practice daily torpor have control of their metabolism at all times. Cold-blooded animals have no such control and cannot raise their body temperature to avoid freezing, but those that are able to move, such as garter snakes in their limestone sink-hole or frogs on the bottom of a pond beneath the ice, are known to relocate to safer places, if they are able, when there is a risk of freezing. Those that cannot move will die if they are not one of the few species that are freeze tolerant.

Metabolism

Metabolism is the chemical process within the body that sustains life. Some substances are broken down from ingested foods to provide energy, others are synthesized internally by the animal. Two main processes are involved: anabolism, which is the biosynthesis of complex organic substances from simpler ones, and catabolism, which is the breakdown of complex substances to release their energy. The metabolic rate is the rate at which warm-blooded mammals and birds use energy to generate heat to maintain their specific body temperature, whether this is their normal temperature or the set point below which they must not drop during hibernation. Entry into torpor is achieved by animals actively lowering their normal set point to the level required for hibernation, and the consequent lowering of heat production allows the animal's temperature to drop, which inhibits the metabolic processes. Endotherms generate heat internally during the process of metabolism and maintain relatively high and constant temperatures that require high energy expenditure. The energy needed by mammals for maintenance is twelve times higher than that required by the ectotherms.

Although there are variations in the physiological changes that occur as animals enter hibernation, the principle is similar. Initially, the metabolic rate slows, heat loss exceeds heat production, and the body temperature drops from the normal mammalian average of 98.6°F (37°C) almost to the point of freezing in many species, a drop of 63°F (35°C), except in the bears and other carnivores where it only drops about 12°F (7°C). The body temperature of animals entering torpor declines either in a smooth curve or in a series of steps.

The temperature set point varies greatly in mammals, but if the environmental temperature drops too low and threatens to freeze them, they shiver, their heart rate speeds up, and their temperature rises sufficiently to avoid being frozen. Few mammals can survive the partial freezing of their bodies. The Arctic ground squirrel is one of them, surviving even when its body temperature drops below freezing, down to 27°F (−3°C) for several days; but it is not frozen solid because of the phenomenon of supercooling, where its body water stays fluid because it contains no objects around which ice crystals can form. Water does not necessarily freeze at 32°F (0°C) despite the common belief; the smaller and purer the water droplets the longer they can remain fluid. The body temperatures of several bats have also been recorded just below the freezing point, in the case of the red bat (*Lasiurus borealis*) down to 23°F (−5°C).

When a mammal enters hibernation heat must radiate from its body to lower its temperature, and the larger the body the longer it takes to cool and then eventually to rewarm before its emergence. The rewarming of a large body back to normal from a temperature close to freezing is a very energy-costly process, far too costly for grizzly bears or polar bears to undertake.

During hibernation the dormant animal's heart rate drops dramatically. The echidna's heart beats only once per minute, the European hedgehog's active rate of 188 per minute drops to 21, and the Alpine marmot's heart rate is only 3 or 4 per minute when it is torpid. The black bear's rate of 45–50 per minute in normal sleep drops to 8–10 when it is dormant, and the chipmunk's very high rate of almost 200 beats per minute drops to just 5 beats. The circulatory system is compartmentalized, restricting the blood flow to the extremities, where the risk of freezing is greatest, and as the body cools down the blood vessels constrict (vasoconstriction) and circulation slows. In the case of the freeze-tolerant species of amphibians and reptiles the blood and all other fluids outside the cells eventually freeze solid.

The respiration rate in warm-blooded animals also slows considerably during torpor. The light-sleeping bear's breathing slows by half, and in the deep-sleeping rodents it drops to two breaths and in the echidna just once per minute; some animals may cease breathing completely, a condition known as apnea, for several minutes at a time. Hedgehogs have survived total immersion in water for twenty-two minutes and noctule bats for sixteen minutes. Marmots and bats have been kept in pure carbon monoxide for four hours and survived without harm. Apnea also frequently occurs in the ectotherms, and for much longer periods. Painted turtles and pond sliders are known to survive without breathing for several months, and the oxygen in their blood can fall to almost zero for four months without causing harm. Amphibians that are frozen almost solid, with just their intracellular water still fluid, are of course unable to breathe. Despite the infrequency of breathing during torpor, in their small hibernacula chamber the dangerous buildup of carbon dioxide does not appear to affect the hibernating animals, whose blood remains well oxygenated during their sleep.

In the hibernating species studied at some length, such as bats, dormice, marmots, and hedgehogs, there is a definite correlation between the rate of respiration and the sleeping animal's temperature. Below 53.6°F (12°C), and certainly when the temperature nears the freezing point, there are long periods (often of several minutes' duration) between breaths. Between 53.6°F (12°C) and 68°F (20°C) the rate of breathing increases and the periods of apnea are progressively shorter, and as the temperature rises above 68°F (20°C) the rate increases until it is normal. At very low hibernation temperatures, such as that experienced by the ground squirrels, breathing occurs as periods of several breaths in sequence (polypnoea) followed by long periods without breathing. In fact, there is considerable proof of apnea in hibernating mammals.

Powering Torpor

Conserving energy by hibernating is an admirable adaptation, but even the low metabolism of torpidity for long periods must be fuelled, in the case of the internal storers without the ingestion of further supplies. So hibernation and estivation are firmly connected to energy and are only possible if adequate reserves are available for the total dormancy. From the physiological aspect, hibernation in the animals that store fat internally has been likened to slow starvation, but starvation is a state of extreme hunger due to the lack of essential nutrients over a long period, and this is certainly not the case during torpidity. Hibernating animals are merely depleting fat reserves that have been calculated to provide energy for their long sleep, arousal, and in many cases for the reproductive season which immediately follows. It is a well-known fact that brown and black bears do not start hibernating until they have stored sufficient food for their sleep. The mammals that store food in their larders can eat whenever they awaken, so they do not starve either.

Fats, not carbohydrates, are the sole source of energy for hibernating mammals.[7] To provide the energy for their long sleep they must store sufficient food in the form of fatty substances or lipids (fatty acids, cholesterol, and triglycerides) in the blood and body tissues. These are oxydized (combined with oxygen) to provide energy to fuel the animal's lowered metabolism, and CO_2 is exhaled. Lipids yield 9 kcal[8] of energy per gram, whereas protein and carbohydrate yield only 4 kcal per gram. Body weight would therefore increase dramatically if glycogen (a carbohydrate, and the storage form of glucose) replaced lipids as the source of hibernation energy. Mammals can live off their stored body fat for long periods, the record probably being the nine months' annual sleep of the Arctic ground squirrel. Animals enter deep hibernation on an empty stomach as digestion ceases when the temperature drops, and unused food could deteriorate and affect the animal's health. Food introduced artificially into a hibernating mammal's stomach has remained undigested.

Hibernating mammals rely upon two kinds of fat to provide their energy. The most plentiful form is ordinary adipose tissue or white fat, which is burned to power the physiological processes. They also have brown fat which is composed of special cells that produce heat in response to changes in the environmental temperature, and therefore have a most important role in providing the energy needed during arousal from torpor. The brown fat cell's cytoplasm (the cell contents excluding the nucleus) is rich in the small structures or organelles called mitochondria (which convert nutrients into energy and which account for the fat's brown color), and in vacuoles, which are the cell's storage units for food, water, and metabolic wastes. Brown fat cells release their energy as heat to raise the body

Black-tailed Prairie Dog *Replete with fat which provides the energy to keep it warm underground for its 4–6 month hibernation on the northern prairies, a black-tailed prairie dog awaits the triggers which will prompt it to commence its long sleep. A facultative hibernator, its dormancy is induced by shortage of food and lowering temperatures, so its starting time is optional.*
Photo: Clive Roots

temperature by the process known as thermogenesis, during which fat molecules are broken down into fatty acids that are then broken down further by the mitochondria and their energy released as heat. Mammals have brown fat at birth, but this is lost as they mature, except in hibernating species such as bats and some rodents. Birds do not have brown fat and the hibernating poorwill and the species that undergo daily torpidity, such as the hummingbirds, swifts, and nighthawks, all rely upon white fat for their energy.

The hibernating amphibian's reduced metabolism uses little energy, but its source of power comes from the oxydation of lipids. Studies have shown that frogs store large amounts of lipids in their bodies during the late summer and fall, and when they emerge from hibernation there is only sufficient left for mating and spawning, until they can feed again.

When animals have prepared their fat reserves it is very important that they use them. The lifespan of captive animals has been shortened when they were not allowed to hibernate, probably as a result of obesity resulting from continuing to eat during the season when they should have been sleeping and using their stored fat.

Freeze Protection

Most ectotherms of northern environments hibernate where they are protected from the low air temperatures of winter. Many become dormant underwater, and others deep underground where they are safe from temperatures that may dip to −40°F (−40°C) above ground, and even lower with added windchill. Some sleep under a shallow covering of soil or leaves, and the snow then provides extra insulation, but in very hard winters with poor snowpack their survival chances would be very low, so they have evolved a unique means of reducing the risk. The freezing point of the body fluids of most land vertebrates is 31°F (−0.5°C), and total freezing has always been thought to have fatal consequences, although researchers recently claimed to have frozen laboratory rats down to 19.4°F (−7°C), and then revived them without ill effects. Freezing severely damages cells, deforming them and rupturing their walls, or dehydrating them as the ice crystals thaw and in doing so extract the cells' water. Freezing can burst blood vessels and prevent oxygen and nutrients from reaching the organs. It is extremely unlikely that an animal could survive the total freezing of its body fluids other than in its extremities, which may then result in the loss of digits or limbs, ears, or tails. However, body fluids are either extracellular or intracellular, and although the freezing of the cell contents is fatal, some animals have evolved ways of protecting the contents while permitting ice crystals to form extracellularly, because they produce their own anti-freeze compounds and are partially freeze-tolerant.

The other major cold protective strategy of some ectotherms is the process known as supercooling, where the freezing point of a liquid is depressed through the removal of all impurities in the body fluids on which ice crystals would normally begin their development. With few exceptions, mammals cannot survive if their core temperature drops below freezing. The Arctic ground squirrel is one of these exceptions, as it can supercool its body water so that it does not freeze, and several northern bats may also have this ability.

■ DURATION, AROUSAL, AND EMERGENCE

The commencement of hibernation and its length are largely dependent on latitude and therefore day length and temperature. An animal's hibernation period and its range or distribution are generally inseparable, and it is misleading to give an exact general period for the hibernation of a species with a wide latitudinal range, and therefore temperature. From Alaska to Florida, for instance, there will be great variation within such a large area. In the far north the black bear may hibernate for seven months, while in the milder Midwest it may only sleep for four months, and in the Everglades not at all. In Europe the badger hibernates in winter in northern Scandinavia but remains active in England and France except during very severe weather. It is much the same for animals with a wide range of elevation, such as the viviparous lizard, which occurs in lowland Spain where it hibernates for three months, but at 11,500 feet (3,500 m) in the Alps it remains torpid for at least seven months.

Sometimes variations in sleep length appear to be individually based, and may have nothing to do with latitude or altitude. For example, some captive gold-mantled ground squirrels hibernated while others in the same cage remained active, and the sleeping ones were not disturbed by the activity of the others; but this may have simply been a case of artificial conditions creating artificial behavior. Surprisingly, the climatic conditions when animals emerge in spring are often very much worse than when they commenced hibernation in the fall. When gold-mantled ground squirrels started their hibernation at Big Bear Lake in California one fall the air temperature was 41° (5°) to 46.4°F (8°C) and the ground temperature was 46.4° (8°) to 59°F (15°C). When they emerged from hibernation in March the air temperature was slightly lower at 39.2° (4°) to 44.6°F (7°C) and the ground temperature very much colder at 35.5° (2°) to 39.2°F (4°C), so the environmental conditions hardly seemed favorable for emergence. Ground squirrels and marmots frequently have to push their way through snow to reach the surface when they end their hibernation. The possible reasons for emerging at such a seemingly unsuitable time are the circannual rhythms of their internal clocks, over which they have no control, or the fact that their food or fat stores have run out and they are hungry, and of course the breeding season cannot be delayed.

In most north temperate regions animals spend approximately six months in hibernation—from October to March. Within the Arctic Circle this period can be extended to nine months for species such as the Arctic ground squirrel and the Eurasian reptiles—the adder and viviparous lizard. Elsewhere, such as the desert regions of the southwestern United States, droughts may result in the normal periods of estivation and hibernation running continuously so that animals like the spadefoot toads may be torpid for eighteen months. But the dormant hibernators do not sleep solidly for their whole term as once believed, and even those that sleep deeply awaken quite regularly. Why a comatose animal should periodically arouse for several hours, or perhaps a whole day, during its long hibernation is unclear and has been the subject of much speculation. Arousal, which entails heat production, is a drain on an animal's precious stored reserves of fat as it raises its body temperature, and there must be a very worthwhile reason for doing it. Possibly life in such a reduced state—called metabolic depression—damages the animal's nerve cells and affects its ability to develop immune responses as defense against infection and disease. Arousal, even for just a few hours, restores the metabolic processes and perhaps allows the animal to recover from the stress of metabolic depression, giving it an opportunity to really sleep and rejuvenate its body, which does not happen when it is dormant. It is also believed that some animals may need to restart their metabolism in order to excrete waste, and it is possible that physiological imbalances occurring during hibernation, such as the depletion of blood glucose, could be corrected during their brief return to normal temperature.

There is also a safety mechanism that functions if the temperature of the hibernacula drops below a mammal's set point, threatening to freeze it. The hibernating animal arouses from its sleep through shivering and violently contracting its muscles, which increases its heart rate and respiration, and thus its oxygen intake and body temperature. In the case of the external food storers, such as the chipmunk, depleted food reserves in its body are the trigger to arouse and feed and thus replenish its energy from its larder. Northern bats, which hibernate mainly in the constant temperature and humidity of caves, shiver and "rustle" their wings to raise their body temperature as they arouse.

In the early stages of arousal the front of an animal's body—the part containing the essential vital organs, the heart and brain—is heated faster than the posterior and the extremities. Blood is directed to the vital organs and especially to the

Greater Horseshoe Bats *Many northern insectivorous bats hibernate in caves where the temperature and humidity are constant. They may hang in tightly packed clusters (see page 173) or hang singly like these greater horseshoe bats in a cave in Kyrgyzstan, central Asia, which wrap their wings around themselves in the manner of the large tropical fruit bats.*
Photo: Courtesy Andrey Ostapenko

brown fat tissue which occurs along the neck and between the shoulder blades of mammals, indicating its importance in providing heat to power the arousal process. Vasodilation or widening of the blood vessels occurs as mammals arouse from hibernation, and arousal time depends on body size. Large hibernators such as the woodchuck and bobak marmot warm up slowly, at the rate of about 1.8°F (1°C) per ten minutes, whereas tiny creatures like the mountain pygmy possum warm up much faster—at the rate of 1.8°F (1°C) per minute. Heat production warms the body at a much faster rate than the cooling-down process at the start of hibernation and animals therefore arouse more quickly from torpidity than they enter it.

Emergence from hibernation is controlled by the same factors as the entry weeks or months earlier—temperature, photoperiod, or body rhythms. The increasing temperatures of springtime are the major factor initiating the arousal and emergence for many species. Terrestrial amphibians such as the wood frog and spring peeper, which hibernate under leaves, are the first to appear as the snow begins to melt, whereas aquatic hibernators like the leopard frog and bullfrog must await the melting of a large part of their pond's ice cover before awakening. Biological rhythms appear to play a greater role in those that sleep deep underground, below several feet of frozen soil, for it seems unlikely that they are influenced by the slightly warmer temperatures or longer days above ground, and they often emerge to find that there is still considerable snow cover. There is rarely food immediately available for animals when they emerge, and it is important that they have enough internal reserves left to sustain them over this essential part of their life cycle. Food storers return to their underground caches to eat until food is available outside.

Notes

1. Several mammals, including the sloths and the echidnas, have difficulty regulating their already low body temperatures during cold weather.
2. The bats known for years as common pipistrelles were recently recognized as two distinct species, the soprano pipistrelle (*P. pygmaeus*) and the bandit pipistrelle (*P. pipistrellus*).
3. One mammal, the Arctic ground squirrel, is also adapted to survive the partial freezing of its body fluids.
4. In Madagascar, prosimians and tenrecs estivate during the winter to avoid cooler dry weather.
5. Excluding the Arctic ground squirrel, which cannot burrow deeply due to the permafrost, and is then surrounded by frozen soil when the shallow top layer refreezes.
6. The processes and functions of an organism.
7. Some northern frogs store glycogen (a carbohydrate and the storage form of glucose) in their livers; it is converted to glucose and functions as a cryoprotectant.
8. A kcal is the amount of heat needed to raise 1 kg of water by 1°C.

2 Sleeping Fish

We rarely think of fish sleeping even daily, let alone for long periods, but when northern rivers, lakes, and even garden ponds are covered with ice and the temperature of the remaining water is close to freezing, fish hibernate just as surely as the other ectotherms—the amphibians and turtles—that occupy the same habitat. An even less-well-known fact is that a number of fish in warmer regions estivate, cocooned in the baked mud of their ponds during the dry season. Hibernation in fish is triggered by temperature, and they react to lowering water temperatures by reducing their activity. In preparation for this they have stored fat to provide the energy, albeit considerably reduced, which they need to support them through several months of inactivity when they are not feeding. Fish in garden ponds are normally overfed in late summer to allow them to build up fat reserves for the winter, when they are not fed at all as their lowered metabolism compromises their digestion.

Hibernation is a dangerous time for any animal that cannot defend itself or escape, and fish resting exposed on the bottom of their pond in a comatose state are no exception. Although carnivorous fish would be similarly sluggish and unable to take advantage of their sleeping prey, they are vulnerable to otters, water rats, and muskrats. However, a greater risk to the survival of sleeping fish may come from their water. In small bodies of water, and certainly in small garden ponds in cold climates, there is a risk of a buildup of ammonia, which can make the water toxic and lethal to fish. Although some ammonia comes from the fish themselves in the form of their bodily excretions and respiration, most is produced from decaying plant life. As fish breathe—even the greatly reduced breathing during hibernation—they exhale ammonia and carbon dioxide (CO_2). The ammonia settles near the bottom of the pond, where the fish are hibernating, and the CO_2 rises to the surface. A supply of oxygen is necessary to counter this buildup, which in naturally still ponds must await the thaw, but in ornamental goldfish and koi ponds artificial aeration is often used. As ice prevents the exchange of gases between the air and the water, allowing the buildup of toxic gases, a hole in the ice must be made to allow these to escape. With the risk of toxic gas buildup alleviated, cold-water fish can remain beneath the ice without harm, and there are records of many species hibernating when their water temperature dropped close to freezing, including carp, tench, roach, and catfish. They become inactive when it drops below 46.4°F (8°C) and rest on the bottom until the temperature rises.

Unlike most materials, in which the density increases as they get colder, water does this only down to 39.2°F (4°C). Between then and freezing it gets less dense and rises to the surface, where it will freeze if the temperature drops to 32°F (0°C). On the bottom, in deep ponds, the water is therefore several degrees warmer than at the top under the ice, and fish can rest there to take advantage of the slightly warmer water. For this reason circulating the water in winter in ornamental ponds, as opposed to simply aerating it, is not recommended. One disadvantage of lying on the bottom of their pond for so long while hibernating, however, is that they collect parasites on their stomachs.

In addition to the fish that hibernate when their water chills, there are also those that have evolved to avoid environmentally unsuitable conditions of the totally opposite kind—when their ponds dry up. Estivation in fish is triggered by drought, and several species of fish have evolved unusual survival behavior for when their ponds and rivers begin to shrink. The lungfish are the acknowledged experts at this, able to survive for several years baked into the mud at the bottom of their pond. A few other fish also avoid the dry season in similar fashion, including the climbing perch of the *Anabantidae* family, which has developed a secondary breathing apparatus allowing it to live out of water; and the airbreathing catfish of the family *Clariidae*. These have a similar labyrinthine chamber over the gills that can absorb atmospheric oxygen, and include the infamous walking catfish (*Clarius batrachus*), now well established in Florida, which usually moves across land when its pond dries, but can submerge in the mud and await the rains.

The most unusual form of estivation occurs in several species of killifish (*Aplocheilidae*), small but very beautiful fish whose eggs combine estivation and incubation for several months. As there are annual plants, which grow only from seed each spring, so there are annual vertebrates—several species within the killifish group. These fish live in temporary seasonal ponds in Africa and South America, and spawn as their pools begin to dry out at the end of the wet season, pushing their eggs into the pond-bottom mud. The adults then die and the eggs estivate for 4–6 months during which time the embryo also develops, so they are ready to hatch when the rains begin. They must then grow and become sexually mature and ready to lay their eggs before their ponds dry out again. The blue-finned notho (*Nothobranchius rachovii*) is one of these egg-estivators; they live in seasonal pans in Kruger National Park and in pools in neighboring Mozambique, which fill only during the rainy season. Another species is the blue notho (*Nothobranchius patrizii*), which has a lifespan of about six months in seasonal pools on the coastal plain of Kenya and Somalia.

■ HIBERNATORS

Some of the Species

Carp (*Cyprinus carpio*)

The natural range of the wild carp is eastern Asia, especially Persia and the Caspian Sea, but it has been widely introduced and now occurs ferally throughout the world, as the original releases were all of domesticated carp. The wild carp is a heavy-bodied fish with two barbels on each side of its upper jaws, and is usually olive-green with a yellowish belly and red caudal and anal fins. It is long-lived, the record lifespan being almost fifty years, and at that age may weigh 75 pounds (34 kg). A toothless fish with a slightly protruding upper jaw, its skin is partially scaled, except in the leather carp which is scale-less.

Carp are omnivorous and eat virtually anything animal or vegetable, but they favor plants and are very messy eaters that destroy aquatic vegetation and silt up rivers with the resultant sediment. They become inactive in winter when the water temperature drops below 46.4°F (8°C), and remain sluggish at the bottom of the pond until the temperature rises. They do not eat during their hibernation; their metabolism slows down and it is believed they may even withstand temporary freezing.

Carp reached China long ago from Persia, and may have already been domesticated to have survived such long-distance overland trade, although the first records of their domestication appear in China in the third century A.D. They were introduced into Europe in the Middle Ages and were initially kept in monastery ponds. They reached Japan in the seventeenth century and were originally kept in the rice paddies as a source of food. Eventually, through selective breeding, they gave rise to the colorful koi carp (see color insert) and the ghost carp, which is a hybrid between the koi and mirror carp. These ornamental breeds rival goldfish as the most widely kept cold-water ornamental pond fish, and are usually wintered outside in northern ponds. Carp are now one of the most widely distributed fish in North America, and have been released purposely in many countries including Australia, Canada, and the United States (in 1877), where they have thrived and are well established; they have become the dominant fish in many inland waters where they are prized as sport fish.

Goldfish (*Crassius auratus*)

The wild goldfish is actually a silvery color, and is a member of the minnow family (*Cyprinidae*) and therefore a close, but smaller, relative of the carp, with which it can interbreed. It reaches a maximum weight of 6½ pounds (3 kg) and is 18 inches (45 cm) long, and lacks the barbels that the carp have on their upper jaws. Its native habitat is China, where it lives in still or slow-moving fresh water, and is a very hardy species, able to survive in water just 1.8°F (1°C) above freezing. It has large eyes and a good sense of smell and hearing. The goldfish is omnivorous, and eats a variety of plant matter, insect larvae, and snails; it is a very prolific fish that lays large batches of eggs at intervals of just over one week, which total several thousand in a season.

The domesticated goldfish is a favorite outdoor ornamental pond fish, even more popular than the koi and other domesticated carp, but it is most well known as a cold-water indoor aquarium fish; it is also available in many mutations of both color and form. There are now goldfish in many shades of yellow, orange, red, and brown, and even black ones. The unusual forms that have been selectively bred include the black moor, which has a double tail and telescope eyes, and the white oranda, which has a large, red "lion's" head.

The domesticated goldfish has been introduced and is now established in many countries, often as the result of commercial breeders releasing unwanted fish, but unlike carp they are considered pests in most places as they are detrimental to the native fish. Like most domesticated animals that become feral and established in the wild, even in countries far removed from their original habitat, they eventually reverted to their natural silver, which has greater survival value than the bright golden color. The goldfish now thrives in many ponds and lakes in the United States, in England, New Zealand, Australia, and throughout Europe.

Tench (*Tinca tinca*)

The tench is a common fish in lakes and ponds in the British Isles, Europe, and temperate Asia. It is dark olive-green and has rounded fins, small slimy scales, and red eyes. There is also a golden phase with dark-brown spots in the United Kingdom. It is a favorite coarse fish of anglers, a strong fighting species, which reaches a maximum weight of 15 pounds (6.8 kg) and length of 18 inches (45 cm). It has also been introduced widely for sport fishing and is now established in many countries including the United States, New Zealand, and Australia. The tench is a carnivorous bottom-dweller that stirs up the mud to find crustaceans, small mussels, and insect larvae. It prefers still water and can survive in almost stagnant ponds with very low oxygen levels; it lies dormant buried in the mud when the water temperature drops almost to freezing and its pond ices over.

Central Mudminnow (*Umbra limni*)

This small but robust fish is a common and widespread species of the St. Lawrence River and the Great Lakes, as well as the Upper Missouri and Mississippi River systems, although it avoids fast-flowing water. It is a favorite bait fish of fishermen, however, despite its name it is not actually a minnow but a small relative of the pike, with a maximum length of about 5 inches (13 cm). Mottled olive-brown above, it has faint dark-brown vertical bars on the sides and a yellowish-white belly, and a short and blunt snout. It may live for eight years, a long lifespan for a small fish, and its major habitat requirement is a thick bottom layer of mud. The mudminnow is carnivorous and eats small snails and crustaceans, insects, small fish, and fish eggs. It tolerates very harsh conditions including low oxygen levels and cold temperatures, and survives in winter by burrowing into the mud tail-first and becoming dormant. It also adopts the same tactics during occasional midsummer dry periods, and is the only fish known to hibernate and estivate.

■ ESTIVATORS

Some of the Species

West Australian Salamander Fish (*Lepidogalaxias salamandroides*)

The salamander fish is a small, elongated species just over 2 inches (5 cm) long, which lives in shallow, highly acidic black-water pools in the heathland peat flats of southwestern Australia between the Blackwood and Kent Rivers. It is a monotypic species, the sole member of its family *Lepidogalaxiidae*, a very unusual fish of uncertain origin and relationship to the other Australian freshwater species. It has an erect dorsal fin and a very flexible neck, which allows it to bend its head up, down, and sideways; this compensates for its lack of eye muscles and the inability to move its eyes. Its body is also very flexible due to the absence of several ribs and the widely spaced vertebrae, allowing it to wriggle through the sand. When the salamander fish's pond dries out in the Australian summer it survives dehydration by burrowing into the sandy bottom of its pool, where it is kept moist by groundwater. Gas exchange—carbon dioxide out and oxygen in—occurs through its skin while estivating, and it stores urea for the whole period of the dormancy, eventually excreting it when aroused by the autumn rains; it then emerges within a few minutes of its pond beginning to fill. The salamander fish is an insect-eater that stores fat reserves in its belly for the estivation from January to May.

Climbing Perch (*Anabas testudinea*)

This perch is a member of the family *Anabantidae* or labyrinth fish, which includes several popular aquarium species such as the paradise fish and Siamese fighting fish, but the climbing perch lacks their color, being grayish-green with a dark spot at the base of the caudal or posterior fin and another behind the gills. It lives in Southeast Asian waters, especially in the Mekong River, where it reaches a length of 10 inches (25 cm). It has a labyrinthine chamber over the gills, which is very rich in blood vessels and can absorb atmospheric oxygen, thus allowing it to breath air when out of the water. A hardy fish, it can survive very unfavorable conditions such as low oxygen, high water temperature, and the drying out of its pools. Despite its name, however, it does not climb trees, but it can walk overland in search of another pool if its pond dries out in midsummer, and it can survive out of water for several days. The climbing perch's body is supported on the edges of its gill plates when it walks, and it propels itself by its fins and tail. Alternatively, to escape hot and dry conditions it may burrow into the moist mud and await the rains, but it is unknown whether it forms a cocoon in order to survive the complete desiccation of the mud or if it needs moist substrate around it to survive.

Lungfish

Lungfish are very primitive animals, relics of ancient fish that were related to the ancestors of the land vertebrates. Their fossils date back to the Lower Devonian Period (380 million years ago) from China, and they are the closest living relatives of the tetrapods (the first vertebrates to walk on land), whom they resemble by having enamel on their teeth, similar skull bones, and a comparable blood system in which the pulmonary blood is separated from the body blood. They also have four limb-like appendages, slender fins that are homologous to the legs of terrestrial vertebrates. The lungfish have internal nostrils that connect the nasal region to the roof of the mouth, another tetrapod characteristic. They are heavy-bodied and powerful eel-like carnivores, and are very aggressive and fearless hunters. They are all obligate air-breathers and their lungs are similar to those of the ancient amphibians, being modified swim bladders, which most fish use for buoyancy and which in the lungfish absorb oxygen and remove carbon dioxide. They all have double or paired lungs, except the Australian lungfish, which has a single one.

There are six species of lungfish, four in Africa, and one each in South America and Australia. With the exception of the Australian lungfish, which cannot survive a long drought, they estivate to escape desiccation during the dry season. The South American lungfish can remain dormant for 2–3 months, but the African species can lie comatose for up to three years.

Estivation in the lungfish involves burrowing into the mud and producing a waterproof cocoon of mucus that dries hard. They become torpid and undergo various metabolic changes, including decreased consumption of oxygen and lowered heart rate and blood pressure. During estivation the lungfish's gills are covered and bound together by mucus to prevent water loss through evaporation, and the lungs become enlarged and have thicker walls rich in blood vessels. They emerge quickly when moisture reaches them at the start of the rainy season and they spawn in their mudhole. In central Africa they are not totally secure in their cocoons, for the shoebill stork has learned how to hook them out of the mud with its large bill. Young lungfish breathe through external gills, like the axolotl, but these drop off as they mature and the adults breathe air. The Australian lungfish (*Neoceratodus forsteri*) is a more primitive fish, the only species in the family *Ceratodidae*. It normally breathes through gills in the usual fish fashion, but uses its single lung when its pond has virtually dried or the water is stagnant, and it neither burrows nor estivates.

African Lungfish (*Protopterus annectens*)

This species lives in the ponds and rivers of West and Central Africa. It is a very large and aggressive eel-like fish that reaches a length of 6 feet (1.8 m) and is totally carnivorous, preying upon other fish and snails which it crushes with its powerful

African Lungfish *With lungs that absorb oxygen and remove carbon dioxide, the African lungfish can survive out of water for up to 3 years when its pond dries out. This long period of estivation is only possible if it can remain moist, which it does in a water-proof cocoon of mucus that dries hard on the outside.*
Photo: Courtesy JT Biohistory Research Hall, Osaka

jaws. Its fins are long and thin and serve as sense organs, and toward the end of the wet season it feeds heavily and lays down a store of fat. As its water dries up it burrows straight down, then turns in a U-shape so that its mouth is upward near the surface. It then secretes large amounts of mucus, which dries to form a leathery cocoon as the mud hardens around it, and it breathes through its mouth via a tube of mucus which extends to the surface. With greatly diminished metabolism it can stay cocooned until the next rainfall, waiting three years if necessary.

South American Lungfish (*Lepidosiren paradoxa*)

This lungfish lives in the central part of South America in oxygen-poor water in swampy areas and in seasonally flooded grasslands. Like the African species it is an obligate air-breather, as its gills have degenerated and it is totally reliant upon its paired lungs, which are connected to the esophagus; it must therefore swim up to the surface to gulp air. The young have gills but lose them as they mature. It has the typical elongate eel-like body of the lungfish and grows to about 3 feet 3 inches (1 m) long. *Lepidosiren* constructs a "nest burrow" for its eggs, which the males protect until they hatch. It estivates during the dry season when the pools of the caatinga—the semiarid scrub forest of northeastern Brazil—and the seasonal waters of the neighboring Pantanal grasslands dry out, burrowing down about 18 inches (50 cm) in the drying mud.

3 Out Cold

Amphibians evolved from fish in the late Devonian Period about 360 million years ago (mya), and then during the Permian Period (290–248 mya) some of the early forms diverged to become reptiles and the modern amphibians. They are therefore the most primitive land-dwelling vertebrates, which form a living link between their gill-breathing ancestors and the lung-breathing reptiles, sharing several of the physiological characteristics of both. Most amphibians are biphasic, meaning they have both a terrestrial and an aquatic phase at some time in their lives. As in all vertebrates, life depends upon oxygen and to acquire it most amphibians employ both gills and lungs. In their larval stages they use gills to extract oxygen from the water, and some even retain gills for use when adult. In addition they breathe through their skins, which have a rich supply of blood vessels and absorb oxygen from the water or the air. With the most complex skins of any living creature, amphibians are becoming increasingly interesting to science because of the many chemical compounds that have potential value in human medicine. Like their fish ancestors and their descendants the reptiles, the amphibians are ectotherms or cold-blooded animals, which cannot generate heat and control their body temperature like the endothermic mammals and birds and are therefore totally dependent upon the environment for their warmth. They cannot keep warmer or cooler than their surroundings, and for most cold-blooded animals it is either too hot or too cold and they must always be environmentally selective. For the temperate-zone species there are basically two seasons, one when they are active, and then their inactive one when they are hibernating, although in some species there is also a third season—when they estivate. During their active season they must continually seek out the most suitable conditions for maintaining their optimal temperature.

It is rarely too wet for amphibians,[1] for they are either purely aquatic and spend all their lives in water or they live in moist soil or in damp habitats usually close to water (even though some cannot swim), and moisture is important to them all. However, it can certainly be too dry, and to maintain a moist skin the amphibians must avoid the desiccating effects of hot and dry weather. They cannot stay out in the sun too long for fear of dehydration, and are therefore more dependent upon the warmth of sun-warmed rocks and soil to raise their body temperature, and consequently many are nocturnal. It can also be too cold, both for activity and the fact that a moist skin must be protected from cold, as it freezes more quickly than a dry skin. They become sluggish when cold and cannot feed, and are therefore excluded from Antarctica and the high Arctic. Most amphibians are dependent on water for breeding and for the development of their larvae, but many species lay their eggs on land and their unique methods of keeping them moist include making bubble nests and even urinating on them. They are freshwater creatures, and although some can tolerate brackish water, none can live in seawater as osmotic pressure would draw the body fluids through their semi-permeable skins.

Despite these important considerations amphibians have colonized some of the world's harshest regions both in the northern temperate zone and in the tropics, and have developed behavioral and physiological adaptations that protect them in subfreezing temperatures and in the midsummer heat of the desert. They have colonized cold-temperate zones due to their ability to escape the low temperatures of winter through hibernating, and they live in some of the world's hottest and driest places as they are able to estivate. Although it seems an unlikely habitat for them, some amphibians do live in deserts and in both temperate and tropical grasslands, which are baked dry by the intense summer sun; they then hide in the soil wrapped in their own waterproof cocoon. Some live at high altitudes in cold mountain streams, an example being *Atelopus ignescens* (which may now be extinct), which evolved on the paramo (the tree-less alpine plateaus) up to 13,750 feet (4,200 m) in the northern Ecuadorian Andes, where the air temperature averages 45°F (7.2°C) and the water is chilly snow melt from nearby mountain tops. Although most amphibians hide from the cold, a few species can withstand prolonged exposure to subfreezing temperatures at high latitudes in the Northern Hemisphere, and can survive being partially frozen. This ability to freeze without harm involves complex metabolic processes and the synthesis of cryoprotectants or anti-freeze compounds that allow them to control which parts of their body can safely freeze. Some species have to contend with unsuitable conditions in summer as well as winter, and then estivate and hibernate in the same year, being active in total annually for only a few weeks.

Survival in temperate regions is not solely dependent upon an amphibian's ability to hide during unfavorable times or to survive freezing. To perpetuate their kind in their chosen environment the summers must allow sufficient time for the completion of their reproductive cycle—finding a mate, egg-laying, incubation, growth of the larvae, and then their metamorphosis into juvenile frogs or salamanders—followed by enough time for them to gain adequate size to withstand the long period of winter dormancy. When all of this is possible they can even live just within the Arctic Circle.[2] Even so, the summers in some regions are too short for them to complete their life cycle and they then overwinter as larvae and complete their metamorphosis the following year. On emerging from hibernation, and usually with sufficient food stores left for the reproductive season, amphibians immediately commence breeding. They lay their eggs in water or in moist places, and the larvae of most species are aquatic. Like fish eggs, with just a few exceptions frog and toad eggs are fertilized externally,[3] the adult's mating behavior involving the sexual embrace known as amplexus in which a male hangs tightly to a female's back and fertilizes the eggs as they are laid. Amphibians produce anamniotic eggs which are shell-less and lack an amnion—the tissue membrane that expands to form the amniotic sac which encloses and protects the fetus during pregnancy—and must be laid in water to prevent their desiccation.[4] A period of lowered temperatures at some stage of their life cycle is an essential aspect of the amphibians' continued survival, for it is necessary for successful reproduction, and even captive animals are cooled to simulate hibernation (a process herpetologists call brumation) to improve breeding success.

Common Toads *The first act of the northern frogs and toads upon emerging from hibernation, even though it may still be very cold and food is not available, is to begin the process of reproduction. In the breeding embrace called amplexus, the male seizes the female around the waist and fertilizes the eggs externally as they are laid.*
Photo: Steffen Foerster, Shutterstock.com

Unlike baby reptiles, which are tiny replicas of the adults, most young amphibians differ from their parents, as they have external gills and long tails. Their metamorphosis from aquatic larvae to adult-like form, if not size, may take several months and includes the development of legs and the loss of their gills and tails in the frogs. The salamanders grow legs but retain their tails into adulthood, and in some aquatic species their gills also. In a few species the eggs are cared for by the parents, and even carried on their backs; and for some the larval development is completed inside the eggs and the hatchlings then resemble the adults.

Amphibians are highly regarded as research animals, to the extent that several species, such as the clawed toad, axolotl, Spanish salamander, Japanese newt, and the tiger salamander, have been bred in the laboratory for so long they must be considered domesticated. They have been used for testing human pregnancy, for the analysis of gonadotrophin, investigating regeneration, and for experimental embryological research. Wild frogs are still collected in several countries, notably India and Dominica, for the gastronomy trade, and in the United States bullfrogs have been farmed for that purpose also. It is unfortunate that amphibian-keeping as a hobby lags well behind the other vertebrates, especially the reptiles, as they are experiencing the greatest decline of all due to their sensitive nature.

Amphibians worldwide are in trouble, foretold by the disturbing disappearance of many populations, especially frogs, and the appearance of others with missing or misshapen limbs. The many reasons blamed for this include disease from viral, bacterial, or fungal pathogens; pollution—primarily acid rain and the widespread use of pesticides and herbicides; and the loss of habitat from wetland drainage, river silting, forest clearance, and urban sprawl. Indiscriminate killing and commercial harvesting for research and for food have also taken their toll; and in addition to all these unnatural pressures, they continue to be a major source of food for a wide variety of vertebrates. The very nature of amphibians, at least their physiology, is a major reason for their sensitivity to change, as they have permeable skins through which there is an exchange of gases and moisture, thus increasing their susceptibility to toxins. But in most cases a combination of factors is responsible for their decline, and although the loss of a population may be blamed upon the draining of a marsh, it is rarely possible to pinpoint a single reason for the general decline of a species. As usual, reconciling the interests of commerce and human progress with the needs of frogs or salamanders is obviously difficult, especially when the reasons for their decline are so tenuous, and may, as in the case of acid rain, result from pollution several countries away. Unfortunately, amphibian populations can crash very quickly.

Within the class *Amphibia* there are three orders. The *Gymnophiona* contains legless creatures resembling earthworms, but none are known to become torpid. The *Urodela* (or *Caudata*) is the order containing the newts and salamanders; and the many species of frogs and toads are included in the order *Anura* (or *Salientia*), which is by far the largest. Many members of these orders hibernate or estivate, and a few actually do both.

■ HIBERNATION

Hibernation is a phenomenon of amphibians living in temperate regions where the low ambient temperature of winter is inadequate to raise their own body temperature sufficiently to allow activity, and their invertebrate prey is probably no longer available anyway. They cannot survive these conditions unless they protect themselves, and there are two ways they do this, which are known respectively as freeze-avoidance and freeze-tolerance. The most obvious means of surviving in winter is to avoid being frozen, and most amphibians adopt this strategy—by finding a place where the frost does not penetrate. Temperate-climate amphibians hibernate underwater beneath the ice, underground in burrows, or on the ground under leaf litter which is then covered by a blanket of snow. But a few leaves and a layer of snow cannot protect frogs in Siberia, Alaska, or on the Canadian prairies from temperatures that may reach −40°F (−40°C), and in most winters they simply cannot avoid being frozen, so these ground-level hibernators have evolved the ability to tolerate being frozen. While many northern amphibians can survive short periods of freezing temperatures such as overnight frosts, these cold-hardy northern species have evolved physiological adaptations that allow them to withstand the freezing of some of their body fluids for much longer periods, without harm.

The most effective way to avoid being frozen is to stay in the water all winter, which is exactly where the aquatic frogs and salamanders spend their winters. Water provides the most stable environment for amphibians, protecting them from the risk of freezing and dehydration, both of which can occur when hibernating on land. The disadvantages are that there is always the risk of ice formation reaching the bottom of relatively shallow ponds during a very severe winter, and the hibernators are trapped until the ice breaks up in spring. This is much later than when the ground-hibernators warm up, and therefore considerably shortens their life out of water and their breeding season. At high latitudes in the Northern Hemisphere several species of frogs hibernate for almost nine months due to the ice cover on their ponds and lakes; although recent studies suggest that they may not just lie comatose on the bottom, even at temperatures close to freezing, but may not greatly lower their metabolic rates and may even be relatively active. The lowering of the metabolic rate of non-freeze-tolerant species is normally triggered by low temperatures. Aquatic species such as the leopard frog and bullfrog normally hibernate underwater, but do not bury themselves in the oxygen-deficient (hypoxic) mud at the bottom of their ponds, as this is a very unsuitable environment. They would probably suffocate there as they need a better supply of oxygen than the mud can provide,[5] and therefore rest on the mud, or perhaps are partially buried, so they can benefit from the oxygen-rich water around them.

Amphibians that hibernate underwater must still acquire oxygen to survive, and gas exchange—oxygen absorption and carbon dioxide excretion—occurs through, or across as it is usually termed, their skins. Northern frogs that hibernate underwater and may be trapped beneath the ice for almost nine months include the common frog (*Rana temporaria*), which is the northern-most ranging amphibian in Eurasia, occurring up to 65°N in northern Finland and Russia (just short of the Arctic Circle); and the northern leopard frog (*Rana pipiens*), which ranges north to Great Slave Lake in Canada's Northwest Territories. Although not northern, but just a high-altitude species, the mountain yellow-legged frog (*Rana muscosa*) of the Sierra Nevada may not breed until July when its ponds are free of ice, but its larvae do not have time to metamorphose before winter so they overwinter as tadpoles before changing into froglets the following short summer. The red-spotted newt (*Notophthalmus viridescens*), which is aquatic when adult, is quite cold tolerant and has been seen moving about beneath the ice in midwinter.

When amphibians hibernate on land, the safest place is deep underground in moist and sandy soil below the level of frost penetration. Species that do this include the spadefoot toads (*Pelobatidae*) and the true toads such as the American toad (*Bufo americanus*), the European green toad (*B. viridis*), and the common toad (*B. bufo*). They are said to tunnel into soil, but they do not actually make a tunnel, which implies kicking soil out behind them like a ground squirrel or mole. Their burrowing involves twisting their hind legs from side to side in a half-corkscrew motion as they move backward into the soil which falls back in place over them. To survive they must dig below the level of frost penetration, which in the northern prairies could be 6 feet (1.8 m) deep. Terrestrial frogs, which normally only enter water for breeding and are not adapted for burrowing, are forced to spend the winter beneath a layer of leaves and the eventual blanket of snow, where they are very vulnerable. When the snow cover is shallow it does not protect animals that cannot generate their own heat, and their adaptations for survival are then concerned with freeze-tolerance rather than freeze-avoidance. The northern amphibians can only tolerate freezing, however, if the ice crystals in their bodies are restricted to the extracellular fluids such as the blood plasma and urine.

Crouched in a tucked position with their limbs beneath them to reduce the evaporative water loss which occurs through their skins during hibernation, partially freeze-tolerant northern amphibians such as the wood frog, boreal chorus frog, gray tree frog, green tree frog, and spring peeper await the onset of subfreezing conditions. Unlike the warm-blooded mammals, which use muscle contractions to raise their body temperature when they are at risk of freezing, the ectotherms lack the metabolic ability to raise their body temperatures as the frost begins to penetrate, but they have evolved various strategies that have allowed them to colonize very cold regions. The most important of these is their toleration of the partial freezing of their body fluids. They can withstand temperatures down to 17.6°F (−8°C), and the wood frog even lower, possibly down to 10.4°F (−12°C), at which time about 65 percent of its body fluids are frozen. Although they cannot survive for long at such low temperatures, they have lived for two weeks at 27.5°F (−2.5°C). When they are frozen all muscular activity ceases, they have neither heartbeat nor respiration, and the frozen blood cannot circulate and carry oxygen and nutrients to the cells. However, the ice that has formed in their bodies is outside the cells only (extracellular), as the cell contents—the cytoplasm and the nucleus—are protected by cryoprotectants or "anti-freeze," which are small molecules that penetrate cells and limit the amount of water that freezes there, preventing cell damage and the death of the animal. As the northern frogs begin to freeze their protective response is triggered and glycogen in the liver and tissues is converted to glucose and is carried throughout the body in the blood. Wood frogs and spring peepers use glucose but the gray tree frog uses glycerols as natural anti-freeze compounds.[6] These depress the freezing point and prevent intracellular freezing and dehydration, although from outward appearances the frogs seem frozen solid, with opaque eyes.

The livers of freeze-tolerant frogs have higher levels of glycogen than those of non-freeze-tolerant species. Within the liver the process of glycogenolysis converts glycogen (the storage form of glucose) to glucose immediately when the frog's skin is aware of freezing temperatures, and the glucose is then carried by the blood throughout the frog's body into the cells, with the higher concentrations going to the vital organs, until the flow of blood is stopped by freezing. The ice that forms in the fluids surrounding the cells draws water from inside them and leaves the glucose solution, which does not freeze. Freezing begins in the frog's toes, due to the high surface area-to-mass ratio, followed by the limbs, and then spreads throughout the body, with the vital organs freezing last, the heart last of all.

Early warning of the onset of freezing is very important to the frogs. The normal freezing point of an amphibian's body fluids is 31.1°F (−0.5°C), but the partially freeze-tolerant frogs control their freezing and actually begin the process themselves before natural freezing begins. They initiate freezing with special ice-nucleators, which trigger the formation of ice in their bodies before it would start

Wood Frog *The most northern frog in the world, which lives within the Arctic Circle, the wood frog can withstand freezing down to 10.4°F (−12°C), but only because its cell contents remain liquid, protected by glucose that acts as an anti-freeze. It hibernates on land under a shallow covering of leaves and soil, plus a blanket of snow.*
Photo: Alexander M. Omelko, Shutterstock.com

naturally. Ice-nucleators are bacteria and proteins, both on the skin and in the intestine, and it is believed that the skin ice-nucleators initiate the formation of ice crystals. So these amphibians start their own freezing process as their temperature reaches 32°F (0°C), just above the normal freezing point. Ice formation is slower at higher temperatures and this fractional difference gives the frog more time to produce and distribute the cryoprotectant. Freezing is also further slowed by the fact that as a body freezes its temperature rises, through the heat of fusion, which allows the cryoprotectant to be thoroughly distributed throughout the frog's body.

Another strategy that northern frogs have evolved to survive freezing temperatures is called supercooling. Contrary to popular belief the freezing point of water is not necessarily 32°F (0°C) but can be several degrees below if there are no impurities in the liquid upon which ice crystals can form. Northern frogs are able to clear their body fluids of substances that would trigger the formation of ice, and extend their freezing point several degrees below 32°F (0°C). These adaptations are naturally restricted to species in which they have evolved, and a tropical amphibian exposed to the same conditions would simply freeze and die. Despite these measures, however, there is believed to be a large winter kill of hibernating northern frogs.

In spring the frog's heart is the first organ to become active when thawing begins, and the flow of blood then restores breathing ability and muscle activity. The terrestrial ground-hibernators awaken when the sun melts the snow cover and warms the soil and leaves beneath, which occurs long before the thick ice has melted on lakes and ponds and the underwater hibernators have revived. This explains why the ground-hibernators such as the wood frogs and spring peepers begin breeding much earlier than aquatic amphibians like the leopard frog.

■ ESTIVATION

Estivation is summer hibernation, to avoid extremes of heat and dryness. It is a behavioral strategy of terrestrial amphibians living in regions, temperate as well as tropical, that experience very hot and dry seasonal weather, and of some aquatic forms whose ponds dry out in summer. They cannot tolerate environments that would dry their moist skins, and need to protect themselves from desiccating conditions during the summer in northern temperate regions and in seasonally arid regions around the world, such as Australia's outback, Brazil's caatinga and Pantanal, and the savannah and bushveldt of Africa. They must find a location where they are not exposed to the sun and hot wind, and they burrow into moist soil or termites' nests and produce a protective covering to prevent dehydration. A source of water and food is also necessary for their long sleep and they survive by storing water in their bladders and laying down a store of fat prior to estivating.

Water storage is inadequate, however, unless precautions are taken to prevent its loss through evaporation, and estivating amphibians make a protective cocoon around themselves, usually with a combination of mucus and layers of dead skin.

The African bullfrog, which estivates for several months during the hot and dry sub-Saharan summers, avoids dehydration, as the sun bakes the land and dries the pools, by burrowing into the soil and enclosing itself in a cocoon made of layers of its own dead skin. Even with its considerably lowered metabolism it slowly uses its body fluids, and there is probably some loss to the environment also; it is quite shrivelled when it emerges, having lost as much as 50 percent of its body water. Its permeable skin allows it to absorb water very quickly, and it soon regains its original contours. The greater siren has survived for several years in its estivation cocoon without eating and drinking, living on stored fat and reducing its metabolic rate by almost 75 percent.

In South America, dry season dormancy with lowered metabolic rate also occurs in the rococo or cururu toad (*Bufo paracnemis*), which occupies the dry thorn-scrub of northeast Brazil; and in the ornate horned frog (*Ceratophyrs ornatus*) of Argentina and Uruguay. They burrow into the soil and shed their skins several times to form a waterproof cocoon around them, leaving just their nostrils clear for breathing. They arouse and shrug out of their shrouds when rains moisten the surrounding soil. South Africa's rain frog (*Breviceps gibbosus*) of the seasonally arid veldt uses the hardened skin on its hind feet to dig deep into the soil and into termites' nests, coming to the surface at night to breed only after it has rained heavily. The temporary puddles do not last long enough for its tadpoles to complete their development, so they grow and metamorphose within the eggs, hatching as tiny froglets.

■ HIBERNATION AND ESTIVATION

Some temperate-zone species live an even more precarious life, exposed to freezing temperatures in winter and then extreme heat and dryness in midsummer, neither of which are conducive to amphibian activity, so they must hibernate and estivate. The western toads and boreal chorus frogs that live in the Craters of the Moon National Monument in Idaho are forced to hibernate when the temperature drops below freezing in winter, and then to estivate in midsummer when the ground temperature may reach 170°F (76°C). Even on the farmlands of the northern prairies, the plains spadefoot toad may estivate during summer droughts and then hibernate from October to March. Couch's spadefoot toad (*Scaphiophus couchi*), at 3 inches (7.5 cm) long the largest spadefoot, lives in very dry regions of the southwestern United States and adjoining Mexico, and only appears from its subterranean estivation or hibernation after rainfall; it has been known to remain dormant for longer than a year waiting for the rains. There are other variations, one of the strangest being the totally reversed behavior of the California tiger salamander (*Ambystoma californiensis*), which is active in winter and estivates all summer. When its pond begins to dry in spring it moves into a ground squirrel's burrow, at first emerging at night to feed if the humidity is high, then remaining in estivation for the summer. It emerges again when the rains start in the fall and then breeds in ponds in winter.

Some of the Species

Salamanders

Salamanders belong to the order *Urodela*, which actually contains two distinct groups of animals—the newts and salamanders—which are referred to by those names but are also collectively called salamanders or urodelans. Unlike the frogs and toads that occur throughout the world except the polar regions, the urodelans are mostly northern, temperate-zone animals, of both the Old and New Worlds. In general they have long tails and cartilaginous pectoral girdles; and they all have limbs, although in some species these are tiny and useless and are present as forelimbs only. Their vertebrae vary in number from twelve to sixty-two, and their skins fit tightly, unlike the loose-skinned frogs and toads. Most species lay eggs, and fertilization is mainly internal, unlike the frogs and toads in which fertilization is almost exclusively external. The larvae of most species have external gills, and some retain these throughout life, a practice known as neoteny. Others have neither lungs nor gills and breathe only through their skins or mouths.

The species commonly called newts have rougher and drier skins than the salamanders, and never have gills when adult. Most newts are aquatic, although many leave the water after the breeding season and may hibernate on land for several months. The species commonly called salamanders have slimy skins due to the mucus secretions of their skin glands, and may be permanently aquatic or terrestrial in moist environments, the land forms entering water to lay their eggs, following which their larval stages are also aquatic. Most of the lungless salamanders (*Plethodontidae*) are terrestrial all their lives, lay their eggs on land, and their young emerge fully formed and free-living. The salamanders are a far more diverse group than the newts, ranging in size from 2 inches (5 cm) to 6 feet (1.8 m). In appearance they vary from four-legged, heavy-bodied animals to some that resemble eels and have tiny forelimbs only; and some are almost fossorial. In keeping with their name the mole salamanders (*Ambystomidae*) of the New World spend most of the time underground.

Salamanders are as vulnerable as frogs to arid conditions, which would desiccate their skins, but generally can withstand much lower temperatures than the other amphibians and may be active when it is almost freezing. They actually prefer temperatures between 50°F (10°C) and 68°F (20°C), with reproduction usually occurring at the lower levels. They live across northern Asia and in the Rockies in cold mountain streams, and have been found on mountainsides 10,000 feet (3,050 m) above sea level. Some species begin breeding while there is still ice on their pools, and the blue-spotted salamander (*Ambystoma laterale*) of central Canada migrates over the snow and ice in temperatures just above freezing to find open water.

Despite claims by Russian scientists at the Magadan Institute of Northern Biology in 1987 that they had revived an Asian salamander (*Hynobius sp.*) that had been frozen solid for ninety years, having survived temperatures down to −40°F (−40°C), it is generally believed that no vertebrate can withstand the freezing of its cells, at least for any length of time. However, Asian salamanders do have very large livers that store glycogen, which is converted to glucose that acts as a cryoprotectant or anti-freeze; laboratory specimens have been frozen to −40°F (−40°C) for short periods and their cells were protected by the anti-freeze and remained liquid. Asian salamanders hibernate on land beneath tree roots or in crevices but if these places fill with water and freeze and do not thaw during the brief summer it is quite possible for them to be locked in the ice for years. However, despite the very low metabolic processes and energy requirement during their period of freezing, it is impossible for an animal's stores to supply energy for almost a century.

Terrestrial Salamanders

Most of the salamanders are terrestrial, living practically their whole lives on land in moist places and entering the water just to breed. But some rarely have access to water, like the alpine salamander in its high mountain forests, or never enter water like the fire salamander, which cannot even swim. Consequently, they have evolved a breeding system that avoids the need for water, their young being born in a fully metamorphosed state, having bypassed the aquatic larval stage. Many of these salamanders live in regions experiencing extremely cold winters, and must bury themselves in soil and under leaves, or enter deep rock crevices or rodent holes for their winter hibernation. Some of the terrestrial species—called mole salamanders—spend most of their lives underground, but emerge in spring to congregate in pools of melt water for courtship and egg-laying.

Tiger Salamander (*Ambystoma tigrinum*)

One of the mole salamanders, the tiger salamander occurs across most of North America from southern Canada to Mexico, excluding the Great Basin, the southwestern deserts, and Florida. It is the largest New World terrestrial salamander, reaching a total length of 12 inches (30 cm), with many subspecies and variations of color and pattern; it generally has a dark background color of blackish- or greenish-brown with yellowish-brown or olive spots or blotches, and usually a brown-and-yellow belly. The tiger salamander is a stocky animal, with a broad snout and tiny eyes, and is aquatic only when breeding; the adults spend much of their lives underground in rodent holes, migrating in spring to shallow and quiet water to reproduce. Several races are recognized, including the gray tiger salamander of the central Canadian prairies and adjacent United States, which is light to dark olive-brown with small dark spots; the blotched tiger salamander of western North America, which is dark brown with yellow blotches; and the eastern tiger salamander from the other side of the continent, which is dark with olive spots. They have been transported across the continent as fish bait and races are now established beyond their natural range, with hybridizing in some areas.

With such a wide range the tiger salamander has a variable hibernation period. In southern Manitoba and North Dakota, where temperatures may drop to −40°F

Eastern Tiger Salamander *The most widely distributed North American species and the largest in the New World, the tiger salamander hibernates for at least 6 months to escape the harsh winters of the northern United States and southern Canada. One of the terrestrial species, it goes underground for the winter, often using old rodent holes.*
Photo: Courtesy USGS

(−40°C), it hibernates from October to April and migrates to ponds as soon as the ice begins to melt. In the higher elevations and heavy snow cover of Yellowstone National Park it has not appeared until early June in some years. In contrast, in the south of its range in southern Texas and adjacent Mexico the race known as the barred salamander (*Ambystoma tigrinum mavortium*) rarely hibernates and may even breed during the winter.

Marbled Salamander (*Ambystoma opacum*)

A smaller mole salamander, this species grows to just over 4 inches (10 cm) long, and can be found in the eastern and southern United States from Massachusetts to Texas. It is a black salamander with extensive marbling in the form of bands and blotches on its back and sides, which contrast with the black background and black belly. The markings are usually gray in females and white in males. The marbled salamander is less dependent upon moist areas, and is often found on dry hillsides, but moisture is essential for reproduction. The female lays her eggs in a shallow depression and guards them until they hatch, which only occurs when the depression fills with water in spring. The larvae are aquatic and feed at night; at the age of five months they leave the water, lose their gills, and from then on live on land. This salamander spends much of its time under leaves or underground, often in burrows 3 feet (90 cm) deep, which is where it hibernates from November to March in the northern parts of its range.

Fire Salamander (*Salamandra salamandra*)

This spectacular salamander is a native of central and southern Europe, east to the Carpathian Mountains of the Czech Republic, Slovakia, Romania, and Poland, and is also found in North Africa and the Middle East. It is a large and robust animal that reaches a length of 12 inches (30 cm) including its tail, making it the largest terrestrial species. It has well-developed parotoid glands and noxious secretions, but is brightly colored to warn potential predators of its toxicity. Although individuals vary considerably in color they generally have bright-yellow, orange, or reddish spots or stripes on a glossy black background. Fire salamanders live in damp forest, often on mountainsides, and generally never far from water, yet as adults they do not swim. They roam nocturnally following rainfall, looking for worms, slugs, beetles, and millipedes. Their reproduction, which takes place on land, is rather unusual. A male produces a packet of sperm which he deposits onto the ground and then grasps the female and lowers her onto it. Most of the fifteen subspecies give birth to larvae that have limbs and gills, which are deposited in water; but a few produce fully metamorphosed young in the spring following winter hibernation, and therefore, almost one year after mating, and these young salamanders have no dependence at all on water. This species is quite cold-hardy and only hibernates when the temperature is almost at the freezing point. In the northern parts of its European range the fire salamander sleeps in caves or deep crevices from November to March, often using the same hibernacula for several years, while in the Middle East and North Africa it may estivate in midsummer (see color insert).

Alpine Salamander (*Salamandra atra*)

An all-black mountain species, from the European Alps and the ranges of western Yugoslavia and Albania, the alpine salamander is 6½ inches (16 cm) long including its tail. It is nocturnal and lives in moist woodland, hiding beneath stones and logs during daylight. On the forested slopes it rarely has an opportunity to enter water and consequently gives birth to metamorphosed young. This lack of water has resulted in the most unusual reproductive behavior. The female salamander produces a clutch of about thirty eggs but instead of laying them she retains them in her body where they develop. Only two complete their development, however, as they grow by eating the other eggs and larvae, and they eventually metamorphose within the mother and hatch as fully formed salamanders. This development takes two years to complete at lower elevations and three years at 5,000 to 6,000 feet (1,525–1,830 m). Most individuals are glossy black, but there are several subspecies, one of which—the golden alpine salamander (*S. a. aurorae*)—is golden-yellow on its back and tail and on the top of its head. This species hibernates in crevices and caves from October to March, often communally in a traditional hibernacula.

Semiaquatic Salamanders

The salamanders that spend time in the water and on land, in varying degrees, are usually called newts, although the names salamander and newt are often used interchangeably. Newts have rougher and drier skins than the species known as salamanders, and never have gills when adult. They generally leave the water after the breeding season, and may hibernate on land for several months; the red-spotted newt has a land stage that spends up to three years as a totally terrestrial animal. Newts have toxic skin gland secretions and are rarely eaten by predators. Their costal[7] grooves are much less distinct than those of the salamanders, and males develop smooth skins, flattened tails, and a spiky tail or full dorsal crest prior to breeding. None of the newts are large animals, rarely exceeding 12 inches (30 cm) in length. There are a few species in North America but most are found in the temperate regions of Europe and Asia.

Great Crested or Warty Newt (*Trichurus cristatus*)

This is the largest European newt, and the most aquatic, but it has a terrestrial phase, which coincides with its period of hibernation. It is a large, warty-skinned animal, reaching a length of almost 7 inches (18 cm), and has dark grayish-brown sides and back, but with so many darker spots that it appears almost black. There are small white spots on the lower flanks, and the belly is yellow or orange with dark blotches. Males develop a high, spiky dorsal crest and a full tail crest for the breeding season. This species occurs in Great Britain although it is absent from Ireland, and ranges across northern Europe from France to the Urals. The great crested newt spends most of the year in or around ponds and small lakes where it breeds in water, and its larvae are aquatic. It is very carnivorous and cannibalistic, and both the adults and larvae eat tadpoles, worms, and insect larvae, with the adults also preying on froglets, snails, and even small newts. Adults pass their inactive daylight hours hiding under logs, large rocks, and also in burrows. Females are mostly terrestrial, and return to water in spring to lay their eggs, preferring still water with heavy plant life, as the eggs are attached singly to leaves which are then folded over them. This species usually hibernates on land in burrows, under logs, stones, or leaves, and is dormant from October to early March, when it heads for the breeding ponds, preferring deep water during the day and shallows at night for breeding.

Smooth or Common Newt (*Trichurus vulgaris*)

A more terrestrial species about 6 inches (15 cm) long, the common newt has a smooth pale-brown or olive-green skin with two dark stripes on its back, and with a black-spotted orange belly, which is paler in the females. It is sexually dimorphic, and in the breeding season the male has a continuous back crest from its head to the tip of the tail. It is a very common species, widespread in northern

Europe and western Asia from France to Russia. When on land it catches insects and worms with its projectable tongue, but in the water it seizes aquatic insect larvae and tadpoles with its jaws, which have tiny teeth. Out of the water it prefers damp situations in woods and fields, but migrates in the spring to shallow pools to breed. It is nocturnal and the daylight hours are spent under rocks and logs and in leaf litter. The smooth newt hibernates, often in groups, from October to March in the northern parts of its range, although it is quite cold-hardy, and may emerge when the temperature is barely above freezing. Breeding commences almost immediately on emergence from hibernation, lasts until June, and is totally dependent on the availability of water. The eggs are laid on submerged plant leaves and initially the larvae absorb oxygen through their external gills, metamorphosing into air-breathing adults at the age of ten weeks. The female leaves the water after breeding and spends the rest of the year on land.

Japanese Newt (*Triturus pyrrhogaster*)

This newt occurs on the main Japanese islands of Honshu, Shikoku, and Kyushu, but not the northern-most island of Hokkaido. It has been invaluable as a medical research animal for embryological studies, especially in Japan, where self-supporting laboratory populations have been maintained for many years. The Japanese fire-bellied newt, as it is also called, is a very attractive animal with a chocolate-brown or black body and a reddish-orange belly covered with black spots or blotches, which reaches a length of 6½ inches (16 cm). It has distinct parotoid glands and a very rough skin, but when in breeding condition the male's skin becomes smooth and its tail develops a bluish tinge. This newt is mainly aquatic, and lives in still and quiet waters where it lays its eggs singly on water plant leaves and wraps a leaf around them for protection. It leaves the water when it begins to get cold and eventually hibernates for almost four months in moist places beneath stones or in crevices. It is highly carnivorous and in the water eats aquatic invertebrates and tadpoles, plus insects that settle on the water's surface. When on land it forages in the soil and under leaves, where its favorite prey are springtails.

Red-spotted Newt (*Notophthalmus v. viridescens*)

This salamander has a very complex life cycle, which includes both a land stage and an aquatic one. The adults, which are totally aquatic, are olive-green with red spots and are typically newt-like, the males developing wide, laterally compressed and crested tails in the spring breeding season. During the winter they hibernate in the water but are very hardy and can usually be seen moving around beneath the ice. If their ponds dry out, they become terrestrial and salamander-like and change color to a dull reddish-brown. They lay their eggs in water and their larvae are also aquatic with external gills, eventually transforming to a land stage when they resemble the terrestrial salamanders. In moist forest habitat, especially in mountainous regions these juveniles are bright orange-red and are known as red efts (see color insert); they may spend up to three years in this stage, never entering water and hibernating under leaf litter and moist, friable soil. Before they reenter the water their coloring changes to dull brown with red spots and a yellow belly. The red-spotted newt is a native of eastern North America from southern Canada to Alabama and Georgia, and reaches a maximum size of 4 inches (10 cm).

Aquatic Salamanders

A number of large species of salamanders have been loosely grouped together as "giant salamanders," and they are all totally aquatic and nocturnal. Although they have very interesting habits the animals themselves are considered rather grotesque. They are members of four families, the *Proteidae* (mudpuppies and olms), which are permanently larval; the *Amphiumidae* (amphiumas), the *Sirenidae* (sirens), and the *Cryptobranchidae*, two of whose three members live in Eastern Asia. They are the Chinese giant salamander (*Andreas davidianus*), which grows to 6 feet 6 inches (2 m) long and is the world's largest amphibian; and the slightly smaller Japanese giant salamander (*Andreas japonicus*). Both live in cold, fast-flowing mountain streams and have a very low metabolism generally; when the mountains are snowbound in winter and the water temperature even colder, they survive in a comatose state for weeks without eating, which can only be termed hibernation. Two American species are also dormant for varying periods. The two-toed amphiuma estivates and the greater siren estivates and may also hibernate.

Two-toed Amphiuma (*Amphiuma means*)

This amphiuma is the largest New World amphibian, which commonly reaches a length of 30 inches (76 cm) although the record is 45 inches (1.14 m). It is a dark-brown or black eel-like amphibian with a very slimy skin, no external gills, and four tiny and useless limbs, that lives in the freshwaters of the eastern United States from Virginia to Florida. It has powerful jaws and is highly carnivorous; it lies in ambush for its prey in the mud or under pond bottom debris with its head protruding in the manner of the marine wolf eels, and consequently is often called a conger eel. With its size and sharp teeth it is capable of giving a serious bite, second to that of a snapping turtle. Although highly aquatic in the still waters of swamps, sloughs, and ponds it may change location at night by wriggling through swampland. This species of amphiuma does not hibernate, but when its pools dry out in midsummer it burrows deeply into the mud to estivate, forming an almost waterproof cocoon to prevent excess moisture loss, and in this position lies dormant until aroused by the late summer rains.

Greater Siren (*Siren lacertina*)

A long, thick-bodied eel-like salamander of the southeastern United States, from Chesapeake Bay to Florida, the greater siren lives in both clear and turbid shallow water, in ponds, ditches, and rice fields, and reaches a maximum length of 38 inches (96 cm). It is olive-gray with indistinct yellowish or black spots, and has tiny forelimbs only and external gills—indicating a highly aquatic lifestyle. It is an estivator; if its pool dries up it burrows into the mud and seals itself inside a cocoon made from shed skin and the mucus secretions of its skin glands, and awaits the next rains. It can survive for several months in this condition. The siren is highly carnivorous, and eats a variety of small animal life including crustaceans, insects, fish, frogs, and especially molluscs. It is nocturnal and spends the day under rocks or submerged logs or buried into the mud, and yelps softly when it is held. In the northern parts of its range it is said to experience a major decrease in its metabolic rate in cold water, but its degree of activity is unknown so it cannot categorically be called a hibernator.

Chinese Giant Salamander (*Andrias davidianus*)

The largest of all the salamanders, the Chinese giant salamander is a fully aquatic species that reaches a length of 6 feet (1.8 m), although such large individuals are rare nowadays. It is dark brown, black, or greenish in color with irregular darker blotches, and has a very rough and wrinkled skin, which is porous to permit respiration, as it lacks gills. The Chinese giant salamander is nocturnal but has poor vision, with tiny, lidless eyes positioned on top of its large, flat head. It relies more on its senses of smell and touch, via the small paired tubercles alongside the mouth, to locate the fish, other salamanders, snails, and crayfish that form its diet, and which it catches with a quick sideways snap of the mouth. It lives in the oxygen-rich, rocky-bottomed, cold and fast-flowing mountain tributaries of the Pearl, Yellow, and Yangtze rivers of central China, up to 3,280 feet (1,000 m). It is a very-long-lived animal that has reached the age of fifty years, no doubt because of its very low metabolic rate. During the winter, when the cold water almost reaches the freezing point, it lies dormant on the bottom of its stream.

Frogs and Toads

Although scientifically the name "frog" applies only to animals in the family *Ranidae*, and "toad" to species within the family *Bufonidae*, the members of all the other families are commonly called either frogs or toads, and collectively may be called anurans after their order *Anura*. They are all tail-less[8] after they have metamorphosed from tadpoles, and this is the most obvious difference between them and the newts and salamanders. Frogs have smooth, moist skin, are mainly nocturnal, and are dependent upon a wet or moist environment; in addition to both terrestrial and aquatic forms there are also frogs that live in trees. They are more graceful in shape than the squat toads, and have long hind legs for leaping. Toads are stouter, often squat animals, mostly with dry, warty skins and short legs more suitable for hopping, and they are also mainly nocturnal or crepuscular. Some have large parotoid glands located behind the cranial crest, which produce highly toxic secretions that can be lethal to mammals. They are more lethargic than frogs, and also more adaptable, having evolved to cope with varying conditions of humidity as well as temperature. There are purely aquatic species but most are terrestrial animals that hide from the sun beneath rocks and leaves, and only require water for breeding. In all but a few species of anurans the eggs are fertilized externally: their breeding behavior involves the sexual embrace known as amplexus in which the male hangs tightly to the female's back and fertilizes the eggs as they are laid.

All northern anurans must hibernate to survive the winter, but some species are quite hardy and when they emerge from hibernation both the wood frog and the common or European frog breed in water just above freezing. Several toads, which by their nature frequent drier places without regular access to open water or to permanently moist habitats, must also estivate in midsummer to avoid desiccation. In the drier zones of the tropics and subtropics both frogs and toads are forced to estivate to avoid hot dry weather and in some cases cold dry weather. Dry season dormancy is accompanied by lowered metabolism, especially reduced respiration and heart rate, but without much change in their body temperature, unlike those that hibernate.

Typical or True Frogs

These frogs have long legs with pointed toes and webbed hind feet. Their skins are smooth and moist, with dorsolateral folds, and they have large and prominent eardrums. During the breeding season the males grow a dark nuptial pad on the thumb for holding the female during the breeding embrace called amplexus. They have a worldwide distribution with representatives on all continents, and are absent only in the polar regions, the Arabian Peninsula, and parts of South America. They are particularly well represented in Africa. Northern species hibernate to escape harsh winter conditions, sleeping either at the bottom of their ponds or under leaves and a blanket of snow, with specialized physiology to withstand freezing.

Common Frog (*Rana temporaria*)

Britain's most well-known amphibian, this is a brown or grayish frog with variable dark markings on its back and legs, and with yellowish flanks and pale underparts. It is a very powerful leaper and swimmer. Males are slightly smaller than females, which reach a length of 3 inches (8 cm). The common frog lives throughout most of Europe and its range extends to the Arctic Circle in Scandinavia, farther north than any other amphibian. It prefers still waters, and is found in ponds, lakes, ditches, and even large puddles, and in urban and rural gardens. In decline up to the 1970s, it has made a comeback since the proliferation of garden ponds, which have provided a secure sanctuary, but the inbreeding that now occurs in such closed populations has resulted in reduced immunity to infections. It hibernates from October to March in Britain, usually in the water at the bottom of its pond, but may spend the winter on land in the protection from the

Common Frog *The common frog is a European species that ranges farther north than any other amphibian in the Old World, its range extending to the Arctic Circle in Scandinavia. It hibernates on the pond bottom, and by the time the ice has completely melted in the far north it may have only 4 months of active life before winter starts again.*
Photo: Steve MacWilliam, Shutterstock.com

frost afforded by compost heaps. It is nocturnal and hunts on land for slugs, worms, and other invertebrates.

Edible Frog (*Rana esculenta*)

The edible frog is widespread in Europe in lowland areas, with isolated populations in Scandinavia, but is absent from the Iberian Peninsula and the southern Balkans. It has been introduced into England where it is common. The females of this species are 5 inches (12.5 cm) long and the males slightly smaller. It is actually a hybrid between the pool frog (*Rana lessonae*) and the marsh frog (*Rana ridibunda*), and a mating between two edible frogs is usually infertile, but they can produce fertile eggs if mated back to one of the original species. It is bright dark-green with black spots on its body and bars on its legs, and with a narrowish yellow dorsal stripe. It is a very aquatic frog, rarely seen far from water, but spending much of its time in the grass on the banks of rivers and ponds, into which it leaps when disturbed. It catches its food—mostly insects and worms—on land or on the water's surface, but never underwater. Like other members of the family *Ranidae*, it has paired vocal sacs situated at the sides of the throat.

The edible frog's hind legs have been a delicacy since the days of ancient Rome, and are still eaten, especially in France. This frog hibernates underwater for several months in northern and central Europe, and in the more northerly parts of its range the development of the tadpole to metamorphosis (which normally takes twelve weeks) may be delayed by low water temperatures. They then hibernate for the winter, completing their development into tiny frogs the following spring.

American Bullfrog (*Rana catesbiana*)

A large frog, varying in color from dark green through shades of brown to dark gray and black, with a whitish or yellow belly. It was originally a native of the eastern and central United States and southeastern Canada, but is now more widespread through introductions; it also occurs west of the Rockies, and in Europe where it is especially well established in Italy. Like the edible frog it is also important in the gourmet food trade, being farmed for that purpose in the United States, and the introductions elsewhere mostly resulted from escapes from commercial frog farms. The bullfrog is sexually dimorphic, the sexes being clearly distinguished by the size of their tympanic membrane—the external eardrum behind the eye—which is as large as the eye in females, but double the size in males. The bullfrog is a highly aquatic species, preferring the shallows of large bodies of water, especially well-vegetated still lakes and bogs. It is the largest true frog (*Ranidae*) in North America, reaching a maximum weight of 1 pound (454 g) and a length of 8 inches (20 cm), although is typically about 6 inches (15 cm) long. Its regular foods are insects, worms, tadpoles, and salamanders, but it also eats small birds and hatchling turtles and snakes. Males are aggressive and territorial, and wrestle with other males that enter their territory. Bullfrogs hibernate underwater on the bottom of their pond, in the northern parts of their range from late October to early April, but they may not become completely active until the water warms up toward the end of May.

African Bullfrog (*Pyxicephalus adspersus*)

A favorite pet frog for the terrarium, this species is also known as the "pixie" frog after its scientific name. It is the second largest of the world's frogs after the huge goliath frog of central Africa. Males reach a length of 9 inches (23 cm) and weigh up to 2½ pounds (1.1 kg), but the females are only half the male's size, which is an unusual situation in frogs as the females are usually larger. The African bullfrog lives in pools on the open savannah, scrub country, and bushveldt of eastern, central, and southern Africa, often in quite arid regions. When its pool dries it burrows into the soil, digging with the sharp and toughened callus on its inner hind toes, and then spends several months underground, protected from desiccation by a water-proof cocoon of shed skin. It has been determined that a frog cocooned in this manner loses only half the water lost by a frog hibernating un-cocooned. It has a very stout body and broad head and a wide mouth, and although there is considerable variation the usual color is olive-green with darker ridges. The male has a creamy-yellow throat, and the female's throat is white. It is a very aggressive frog, capable of swallowing mice, small lizards, and other frogs, the tooth-like projections in the lower jaw helping to maintain a grip on its struggling prey.

Wood Frog (*Rana sylvatica*)

This is a frog with a wide range of body color including brown, tan, and rust, with a black "mask" extending from its nose through the eye to the tympanum or exposed eardrum, and to the base of the front limbs. Females reach a length of 2¾ inches (7 cm) while the males are smaller. It is the northern-most amphibian in North America, occurring in Alaska and northern Canada—along the shores of Hudson's Bay and in Labrador. It shares with Eurasia's viviparous lizard the title of the world's northern-most poikilothermic or cold-blooded land vertebrate; it is the only frog in the world that lives within the Arctic Circle (to 70°N) and the only amphibian in northern Alaska, whereas the southern part of the state has three frogs and three salamanders. The wood frog can withstand being frozen, possibly to as low as 10.4°F (−12°C), by producing glucose from the glycogen stored in its liver, which acts as an anti-freeze and lowers the freezing point, thus protecting the cells' contents from freezing and rupture, while the fluid between the cells (the interstitial fluid) freezes. It can tolerate extremely high blood sugar levels. The wood frog is a woodland animal, which enters water only for breeding, and hibernates from early October to early April under leaf litter, rocks, or fallen trees. With little protection from the elements it is not only vulnerable to the extreme cold but is also the first to warm up in spring, long before the underwater amphibians arouse, and it therefore breeds early, before the ice has completely melted.

Northern Leopard Frog (*Rana pipiens*)

The leopard frog is the familiar amphibian of biology classrooms and laboratories, a wide-ranging and formerly abundant cold-tolerant species of the northern United States and Canada, as far north as Great Slave Lake, Hudson's Bay, and Labrador. It is about 3½ inches (9 cm) long and has a background color of green or brown with many circular dark "leopard" spots all over its body. Two pattern mutations are also frequently seen, one with spots or speckles on a tan background, the other completely lacking spots. In many parts of its range it has been declining steadily for several decades, for unknown reasons, but with the usual speculation that climate change or acid rain is the likely cause. It is an aquatic frog both in its breeding and hibernating behavior, and spends the winter underwater resting on the mud at the bottom of a pond or lake. It is therefore more protected from the elements than the terrestrial species, as below the ice the water temperature stays just above freezing. In the far north it hibernates from September to May. After the spring breeding period it is quite terrestrial and wanders away from water, spending most of the summer searching for insects and worms in fields or grassy roadside ditches, hence its alternate names of grass frog or meadow frog.

Typical or True Toads

These are medium- to large-sized toads, with plump bodies and short legs and a dry and warty skin. Their faces are round and they have horizontally elliptical pupils and large parotoid glands behind the eyes, which secrete a very toxic substance sufficiently powerful in some species to kill a dog. They also have visible tympanums or external eardrums, but lack outer ears. These amphibians are distributed worldwide except in the polar regions and Australia, and they have not colonized Madagascar nor some oceanic islands. Males are smaller than the females, and several species darken in color in response to temperature changes. As they are terrestrial by nature they also hibernate on land, generally by burrowing deep or entering holes created by rodents or rabbits to escape the frost.

European Green Toad (*Bufo viridis*)

Certainly the most attractive European species, this is a yellowish-white toad mottled with green, with some warts on its back, and more slender and long-limbed than the common toad. Despite its name it occurs also in North Africa and in southern and central Asia. It frequents lowlands in Europe but in Asia can be found high on the mountainsides. Females are a little larger than males and reach a length of 4 inches (10 cm), and the males have nuptial pads on their fingers when in breeding condition, which help them grip the female during amplexus. The green toad prefers dry and sandy habitat, and often frequents developed areas where it hunts insects attracted by street and house lighting, and only enters water in the spring for breeding. It is not a highly active creature, preferring to sit and let its prey—slugs, worms and woodlice—approach within reach. The green toad has few, if any, enemies due to the very toxic contents of its parotoid glands. It hibernates in winter in most of its range, and then estivates in midsummer in the southern latitudes to escape extreme heat and drought. But generally it is a resilient animal, able to counter very hot and dry summers and then subzero winters by digging deep into the soil, and it can tolerate poor-quality water, even when it is quite polluted.

Common Toad (*Bufo bufo*)

The largest European toad, the common toad is a heavy-bodied, warty-skinned animal, with prominent crescent-shaped parotoid glands that secrete powerful toxins. It is a solitary, terrestrial amphibian, preferring dry habitat in forests, fields, gardens, scrub, and marginal grasslands, and is not too dependent upon water except for reproduction, when it congregates in large numbers in shallow ponds. The common toad lays up to 6,000 eggs in strings which may be 6 feet (1.8 m) long, but they and the eventual tadpoles are very vulnerable to predation, and only a small percentage survive. Widely distributed, it ranges over the whole of temperate Eurasia, including the United Kingdom, but is absent from Ireland and many Mediterranean islands. It is uniform brown or olive above often with dark blotches and with a paler belly, and when adult averages 5 inches (13 cm) in length, with the females being considerably larger than the males. It both walks and hops, and is nocturnal, generally hiding each day in the same favorite place. The common toad eats invertebrates—insects, slugs, snails, and worms—but adults have been known to eat toadlets. It hibernates in leaf litter or in the abandoned holes of rodents and rabbits for at least six months at the northern edge of its range.

American Toad (*Bufo americanus*)

This toad, which reaches a maximum length of 3½ inches (9 cm), is very similar to the Canadian toad, but its cranial crests do not join to form a raised bump or boss as they do in that species. It varies considerably in color and may be brown or brick-red and olive, with dark spots that each contain a wart; there is often a light stripe down its back and its belly is paler and heavily spotted. The American toad is a very common amphibian of moist, but not wet, habitat, including forests, fields, and gardens, throughout the eastern half of North America from northern Canada almost to the Gulf Coast in Louisiana. It is nocturnal and hides under rocks and logs or in leaf litter by day, appearing at dusk to search for insects, worms, and slugs which it seizes with its sticky tongue. It rarely enters water except at breeding time, and in spring migrates from its hibernacula to the closest body of water, where the males make long trilling sounds to attract their mates. Hibernating underground, it burrows down several feet in moist but loose soil to get below frost penetration, but it does so rather late, for even in the northern parts of its range it may not disappear until early November, and then stays underground until May. In addition to the toxic secretions of its large, elongated parotoid glands, the American toad also puffs itself up to deter predators, and "plays dead," but this does not fool the raccoon, who turns it over and eats the underside to avoid the glands.

Great Plains Toad (*Bufo cognatus*)

This is a common toad of the American grasslands and the arid Southwest with a large range from southern Canada (Alberta, Saskatchewan, and Manitoba) through the Great Plains into Mexico, and also extending west into Arizona and New Mexico, so it has evolved to survive in regions that experience very hot and dry conditions. It is the only indigenous toad with large, dark blotches, these being dark green or olive-gray in almost symmetrical pairs, bordered with light brown, and they are very warty. This toad is highly nocturnal and burrows at dawn and remains hidden all day, although it may be active by day during wet weather and when breeding. It suffers high mortality in winter, but can produce huge numbers of young when conditions are suitable in summer. The Great Plains toad also burrows down into moist soil or uses rodent holes to escape both high temperatures and low humidity, and stores up to 30 percent of its body weight of water in its bladder for its estivation. During its dormancy its heart rate drops to half the normal rate of about sixty-five beats per minute. It also stores fat for its winter hibernation, and the two periods of dormancy may run consecutively with the toads disappearing before the end of August in the northern prairies and not reappearing until late spring.

Tree Frogs

The family *Hylidae* contains about 700 species of very smooth-skinned frogs with long legs and big toes, and fingers that look and act like suction cups. They are usually bright green with markings on their sides or legs, but like the chameleon can change color, although they do this mostly for reasons of temperature or mood rather than attempts at camouflage. They are mostly small frogs, less than 2½ inches (6 cm) long, and despite their name they are not all arboreal, but all lay their eggs in water—some in tree-top bromeliads. They are found throughout the world, except the polar regions, Madagascar, sub-Saharan Africa, and most of India. Several tree frogs occupy northerly latitudes that experience freezing temperatures, and are able to withstand the freezing of their extracellular body fluids; in hot and arid regions they have evolved adaptations to avoid desiccation by burrowing and storing water. Their skins are generally less moist than those of the typical frogs and the northern species even bask in the sun and can darken their skins to absorb more warmth.

European Tree Frog (*Hyla arborea*)

This is a plump little long-limbed and smooth-skinned frog, just 2 inches (5 cm) long, from central and southern Europe and the northwest coast of Africa. It has been introduced into the United Kingdom on several occasions but has not survived—due either to the coolness of the climate or collectors who supply the pet trade. It is also called the common tree frog and the green tree frog, but other species also have those names. The European tree frog has small, disc-like pads on its fingers and toes, and varies considerably in color, but is usually bright green with a dark stripe from the eye along the flanks to the groin (see color insert); it may also be yellowish-brown to dark brown with black blotches, and the males have a large yellow vocal sac under the chin. When it emerges from hibernation it is usually brownish-gray. The European tree frog lives in well-vegetated habitat usually near the water on which it depends for breeding. It is a good climber, often going high into trees, and although nocturnal it often sunbathes. It seeks its food in the trees, eating invertebrates such as spiders, flies, crane flies, beetles, and butterflies, and it stuffs food into its mouth with its forelimbs. In the southern parts of its range it is active all year, but in central Europe it hibernates from October to March.

Northern Water-holding Frog (*Cyclorana australis*)

This medium-sized, smooth-skinned frog, about 4 inches (10 cm) long, is one of the water-conserving frogs, of which there are several kinds in Australia. Its color varies from olive to brown or pinkish-brown, depending on its age and condition. It lives in the drier regions of northern Australia, and despite being a tree frog it has terrestrial habits and survives droughts by storing fat in its abdomen and a large quantity of water in its enlarged bladder. Then it burrows into the sandy soil and forms a protective, moisture-proof keratinized cocoon of shed skin, to which further layers are added during estivation, and which becomes quite thick, covering the frog except for its nose. With their slow metabolic rate, internal water and fat supply, and protective cocoon, water-holding frogs can survive for years underground if necessary. The aborigines dig them out and use them as a source of water, placing the frog's rear end into their mouths and squeezing the water out. When it rains the estivating frogs dig their way to the surface, and breed in waterholes, creeks, and dams. They eat earthworms, beetles, and flies, and are quite cannibalistic, eating small frogs, including the newly metamorphosed froglets of their own species.

Gray Tree Frog (*Hyla versicolor*)

Despite its name this frog has very variable coloration, and can change like a chameleon to a wide range of grays and greens in response to its temperature and activity level. It has a white belly, its hindlimbs are yellowish-orange beneath, and it has a white spot under the eye. Its skin is very warty and it could be mistaken for a toad, but it has large toe and finger pads which produce mucus and can cling to smooth vertical surfaces. Females are about 2½ inches (6 cm) long, whereas the males are only 1¾ inches (4.5 cm) in length. The gray tree frog is nocturnal and arboreal, and lives in eastern North America from southern Canada down to northern Florida and Texas. It hides on tree trunks during the day, where its coloring matches the bark and provides protection from predators. Although a tree-dweller, it breeds in water and attaches its eggs to submerged leaves. After life as a tadpole for about eight weeks the newly metamorphosed froglets, which are bright green and stay that color until near adulthood, remain close to water until hibernation time so they do not commence their arboreal lifestyle until the following spring. This frog crawls under leaves on the forest floor to hibernate and may freeze despite the added insulation of the snow cover; but like the wood frog, it is one of the few amphibians that can withstand partial freezing (being protected by anti-freeze compounds), unless the temperature drops below 19.4°F (−7°C), when it will certainly freeze and die as its natural protection is not that effective. It is aroused in early April by the rising temperatures, and migrates to ponds of melt water where it mates and lays up to 2,000 eggs in small clusters.

Spadefoot Toads

These small and plump toads with large heads and prominent eyes differ from the true toads (*Bufo*) in several ways. They lack the large parotoid poison glands and tympanums of the true toads and they have only a single, sharp-edged "spade" on their hind feet for digging, whereas the other toads have two rounded tubercles. Also, they have teeth in their upper jaw, whereas the true toads are toothless, and they have vertical, cat-like pupils in contraction, unlike the true toads, in which they are horizontal. Their skins are smooth like a frog's and not warty like most toads. Spadefoots burrow backward into the soil, where they remain all day and during very hot and very cold periods.

Couch's Spadefoot Toad (*Scaphiophus couchi*)

This is the largest spadefoot, reaching a length of 3 inches (8 cm), which has a ground color of bright greenish-yellow with black or dark-green blotches and a

Couch's Spadefoot Toad   *A toad of the short-grass plains and arid lands of the southwest, this spadefoot spends most of the year buried in the sandy soil, estivating to escape desiccation in the summer and then hibernating when the temperature drops below 60°F (15.6°C). It appears within minutes of the summer rains commencing and must feed and breed frenziedly to make use of the temporary pools.*
Photo: Courtesy National Park Service

whitish belly. It is a native of the American Southwest from Oklahoma to Arizona and then south into Mexico, where it lives on the short-grass plains and in the creosote bush desert, where it is adapted for life in a zeric or arid environment. It burrows to avoid cold and hot and dry conditions, using the sickle-shaped "spades" on its hind feet to reach a depth where the soil is moist and frost will not penetrate, but also uses rodent holes. Couch's spadefoot hibernates in all regions where the temperature drops below 60°F (15.6°C), and estivates in midsummer to avoid desiccation, appearing only to breed during summer rainfalls. It may spend most of the year underground, breeding "explosively" during the spring and summer rains, when its invertebrate food—especially beetles, grasshoppers, and termites—may be so plentiful it can eat up to half its body weight at a time, which soon provides sufficient stored energy to last until the next rains. To benefit from the brief rains and presence of pools it has evolved rapid incubation of its eggs—just fifteen hours in water at 86°F (30°C)—and its tadpoles metamorphose into toadlets in ten days. This species has a long, drawn-out call that has been likened to the bleating of a goat or sheep.

Common or European Spadefoot Toad (*Pelobates fuscus*)

A smooth-skinned nocturnal species distributed from France across temperate Eurasia to southern Siberia, this species is primarily terrestrial, but breeds in spring in pools and marshes. It prefers sandy soil, but can be found in loose soil in forests and in freshly tilled fields and gardens. Its distinguishing features are a lump on the top of its head, and a large pale "spade" on each hind foot—actually a metatarsal tubercle—with which it burrows. Reaching a length of 3 inches (8 cm), the common spadefoot toad is variable in color, usually yellowish-brown or gray with dark-brown blotches or stripes and small orange spots on its sides. When threatened it inflates its body and jumps up at the attacker with open mouth. It is a clumsy-looking animal, but is quite agile and a good swimmer, although it generally only enters water during the breeding season, which begins in April when the toads emerge from hibernation, and lasts for two months. Their tadpoles are the largest larvae of all the European anurans, growing up to 6 inches (15 cm) long before metamorphosing into toadlets. Spadefoots burrow down several feet to avoid winter frosts and begin hibernation in the north of their range in October.

Great Basin Spadefoot (*Spea intermontana*)

This is an arid land toad, which lives in the dry grassland and sandy, open woodland of the Great Basin, from southern British Columbia to the Colorado River, west of the Rockies and east of the Sierra Nevada and the Cascade Range, an area of extreme temperatures. It is a small and rotund species, in which the female is 2½ inches (6.5 cm) long and the male a little smaller; it is gray or olive-green with a bump or boss between its eyes, and with dark-brown or reddish warts, a pale belly, and pale stripes along its back and sides. To avoid the cold and very hot and dry periods it may be dormant underground for eight months in combined hibernation and estivation, either having burrowed into the soil itself or entered a rodent burrow. As it is nocturnal it also spends the day underground. Summer rainfall brings it out of estivation and the warming of the soil and its moistening with snow melt brings it to the surface again after its winter hibernation, which lasts from October to early April. It can survive extreme desiccation during these periods and may lose half its body weight, which can be regained very quickly through skin absorption after the next rains. The Great Basin spadefoot breeds in seasonal ponds, which normally dry out by August, so it has evolved rapid development—from egg to metamorphosis in just six weeks. It eats a variety of invertebrates, including crickets, beetles, grasshoppers, and earthworms.

Pipid Toads

The *Pipidae* is an entirely aquatic family of toads, found only in sub-Saharan Africa and in northern South America. Their feet are completely webbed, and like fish they have lateral line organs which are sensitive to motion in the water, and alert the toads to potential prey or predator. They lack tongues, but can make clicking sounds through movement of the cartilage in the larynx, and together with their modified ears can produce and receive sounds underwater. There are thirty species of these "tongue-less toads," most of which live in Africa where hibernation may not be necessary but estivation certainly is, and they have evolved to escape hot and arid conditions that dry up their ponds. The most familiar of the African species are those of the genus *Xenopus* which are common in the pet trade and have been used extensively for many years for medical research.

Clawed Toad (*Xenopus levis*)

The clawed toad, which grows to 4 inches (10 cm) long, is usually brown above with a pale belly, but albino specimens are frequently seen. Females are usually a little larger than males. Characteristics of this species are their habit of floating motionless below the surface for long periods and their energetic amplexus behavior, in which they swim in circles up to the surface where the females lay their eggs. This toad breeds readily and after being used as a major research animal for many years is considered domesticated. Its original value to science was for pregnancy testing, but antibiotics recently discovered in its skin are now being developed. An aquatic frog, it has very long claws which are used for digging invertebrates out of the mud. It rarely leaves the water, but may do so to relocate to another pond, when it moves very clumsily on land, and it avoids drought and the drying up of its water by burrowing deep into the mud and awaiting the rains. It has now become a pest in several countries, since laboratory specimens were released into the wild.

Discoglossid Toads

These are small, primitive toads in the family *Discoglossidae*, which have tiny, rounded disc-like tongues and warty skins that have glands containing a very powerful toxin, although they lack actual parotoid glands. They cannot shoot their tongues out to catch their prey in the normal anuran manner, and they have the primitive feature of an arciferal pelvic girdle, in which the two halves overlap and have independent movement, whereas in many species they are fused at the midline. They also lack a tympanic membrane, and their pupil has a triangular shape unlike all other frogs and toads. The discoglossids occur in Europe, North Africa, and the Middle East, and there are several species in two genera—the fire-bellied toads (*Bombina*) and the midwife toads (*Alytes*). Both groups lay their eggs in water, but from then onward the male midwife toad cares for the eggs. These toads have an unusual lifestyle, spending half their lives in water and the other half on land—most of this period in hibernation.

Fire-bellied Toad (*Bombina bombina*)

The fire-bellied toad is a semiaquatic amphibian from eastern Europe, a toad of the lowland regions and the still waters of ponds, marshes, and lakes. Small and stocky, just 2 inches (5 cm) long, it is a very attractive animal, with a dark, mottled grayish-green back and a brilliant mottled red-and-black belly which advertises its extreme toxicity, and when attacked it bends over to expose its belly to warn predators of its offensive taste. It is quite hardy, active at temperatures down to 50°F (10°C), and unlike most toads it is diurnal as well as nocturnal. The similar-sized yellow-bellied toad (*Bombina variegata*), from central and eastern Europe, is gray or brown above, with a yellow belly mottled with dark gray. It is also highly aquatic, but is less hardy than the eastern one. In both species the females are a little larger and much stouter than the males. The fire-bellied toad spends approximately half of the year, from April to September, in ponds where it breeds and then forages to store energy for winter. It then becomes terrestrial and in October hibernates on land, buried in soft soil, until the following March. Young toadlets spend their first two years on land, then migrate to a pond for breeding at the start of their third year, thereafter following the adult toad's annual routine. The fire-bellied toad has recently been removed from the *Discoglossidae* by some taxonomists and placed in the family *Bombinatoridae*.

Midwife Toad (*Alytes obstetricans*)

A plump and short-legged toad with warty gray skin and vertical pupils, the midwife toad reaches 2 inches (5 cm) in length. A native of western Europe, like the other members of the *Discoglossidae* it is terrestrial and nocturnal; it is a very slow mover that lives in woods and gardens, and hides by day under logs, in crevices, or digs itself into moist, sandy soil. It sings while buried and can be heard above ground. In this species, unlike other frogs and toads that breed in water, mating occurs on land, and then the male acts as midwife. Fertilizing the strings of eggs as they are laid, he wraps them around his hind legs and back and carries them for several weeks before depositing them in water just before they are due to hatch. A related species, the Mallorcan midwife toad (*A. muletensis*), was virtually exterminated before being saved by captive-breeding programs.

Notes

1. From the Greek *amphibios* meaning "having a double life"—living on land and water.

2. In North America the range of the wood frog (*Rana sylvatica*) extends to 70°N—just inside the Arctic Circle.

3. A notable exception is the tailed frog (*Ascaphus truei*), whose "tail" is actually a copulatory organ.

4. Amniotic eggs (such as those laid by the reptiles and birds) have their own internal fluids and can be laid on land without dehydrating.

5. The hibernating turtle's metabolism is so slow it can get sufficient oxygen from the mud, providing its head is exposed as it absorbs oxygen through the mucus membrane of its throat.

6. Urea (the principal waste product of urine) may also be involved as a cryoprotectant in amphibians. It accumulates during hibernation as the animal's body loses a large percentage of its normal water content.

7. Vertical grooves on the sides of the newts and salamanders.

8. The tailed frog (*Ascaphus truei*) has a short, tail-like copulatory organ, and two species of New Zealand's *Leiopelma* frogs have tiny tails when they hatch as froglets from the egg, but these are soon lost.

4 Uncontrolled Hypothermia

In the evolution of the vertebrates the reptiles lie between the lowly amphibians and the more recently evolved birds and mammals, and were the first vertebrates to become adapted to a completely terrestrial way of life. They are believed to have first appeared about 250 million years ago in the Permian Period, and were then the dominant form of life in the following Triassic Period. Reptiles have lungs for breathing air throughout their life cycle and do not have a gill-breathing aquatic stage like most amphibians. They can be distinguished from the higher vertebrate animals by their protective body covering of scales, shields, or plates instead of fur or feathers, but they vary considerably. The tortoises are hard-shelled and high-domed, while some turtles are flat and have soft, leathery shells. Snakes are streamlined and legless, the typical[1] lizards have four legs and whiptails, and the crocodilians are massive and heavily scaled.

Reptile skin is thick, scaly, and pliable, made of alpha keratin on the inner surface and the harder beta keratin on the outside. Only reptiles have this kind of skin, which reduces water loss through evaporation. Snakes and lizards are considered the most modern reptiles, as most of their speciation (the evolutionary process which gives rise to new species) has occurred in the past 60 million years, whereas tortoises have changed little in 100 million years. Like the amphibians, birds, the monotreme mammals (the platypus and echidnas), and most members of the order *Insectivora*, reptiles have a cloaca—the single chamber into which the digestive, urinary, and reproductive systems empty. In many species, especially those of hot environments, most of the water is reabsorbed when wastes from the kidneys enter the cloaca, and semisolid wastes are then excreted. Reptiles, excluding the crocodilians, have a three-chambered heart.

There are four groups or orders within the class *Reptilia*. The tuatara is the only living member of the order *Sphenodontia*, and the lizards and snakes make up the largest order *Squamata*. The tortoises, terrapins, and turtles belong to the order

Testudines, or *Chelonia* as it was formerly called, and are consequently still known collectively as chelonians; the crocodilians belong to the order *Crocodilia*. The tuataras hibernate, as do many lizards and snakes and chelonians, but the crocodilians are mainly animals of the warmer regions of the world. Of the two North American species, the crocodile has limited low-temperature tolerance and is restricted to the southern tip of Florida, but the alligator has a wider range around the Gulf Coast and eastern seaboard to about 36°N latitude in North Carolina, and becomes dormant in the low water temperatures of midwinter. Similarly, its close relative the Chinese alligator, which barely survives in central eastern China north to about 31° latitude, becomes dormant in riverbank burrows from early November to the end of March. These two alligators have the northern-most range of all the crocodilians. The Nile crocodile, whose range also extends into temperate southern Africa, digs into riverbanks and lies dormant during cold weather, large individuals having chambers in which several may pack tightly together. In the very hot and seasonally dry part of their range they may be forced to dig into the riverbed to find moist sand when the river dries up, and thus as a species practices both long-term winter dormancy or hibernation, and summer dormancy or estivation.

Reptiles, like fish and amphibians, are cold-blooded animals or ectotherms,[2] which cannot generate their own heat like the warm-blooded mammals and birds or endotherms.[3] They lack insulation and use external heat sources to raise or lower their body temperature, and are therefore dependent upon their surroundings. They are active when they are warm and sluggish when cold, as their muscle activity depends on chemical reactions that work faster in a warm body. They increase their temperature and their metabolism by lying on warm soil or rocks, or by basking, when they lie perpendicular to the sun to get the full benefit of its rays on their bodies and expand their rib cages to increase their body surface. Some can darken their skins to absorb more heat. When they need to cool down they reverse this behavior, lying parallel to the sun's rays, seeking a shady spot or lightening their skin. But the terms "cold-blooded" and "warm-blooded" are misleading as they imply that an animal's body is either warm or cold. While a high body temperature at all times is certainly true of most nonhibernating mammals,[4] in which a substantial drop in body temperature is fatal, the hibernating species can withstand hypothermia—abnormally low core temperatures—as they have control of their metabolism to prevent freezing. In contrast, the ectotherms regulate their temperatures through environmental selection, and while they are active they may be as warm internally as the warm-blooded birds and mammals, which range from 96.8° to 107.7°F (36° to 42°C), and the body temperature of some desert reptiles during peak activity may be higher than that of the warmest endotherms. The regulation of body temperature (thermoregulation) is achieved through a complex arrangement that involves external heat sources, chemical reactions, hormone production, and behavior. Reptiles do not just assume the air temperature of their environment, but maintain a temperature at which they are most active by alternating their activities, which include basking in the sun (heliothermy) and then seeking shade, absorbing heat from sun-warmed rocks and soil (thigmothermy) and then visiting cool and moist places. Nocturnal reptiles are obviously thigmotherms, since they hide by day, and although some may bask briefly, most rarely, if ever, see the sun, and acquire their body heat from the warmed substrate. Behavioral thermoregulation is very low cost, for heating by the sun does not use energy, nor does crawling into shade to cool down; reptiles do not sweat and thus do not lose valuable fluids. But there may be some cost to thermoregulation, for in many situations it exposes the reptile to potential predators.

The aim of thermoregulation is to achieve the preferred body temperature (PBT), the range favored by a reptile in which it can function most effectively, its body sufficiently heated to move, to acquire and digest food and to reproduce. Most of their behavioral and physiological functions are influenced by temperature, and its regulation is related to the reptile's environment, as this determines the availability of the various micro-niches, which are important, including a safe place to bask and shelter to cool down. But in temperate climates a reptile's environment changes on a daily and seasonal basis, and it must regularly adjust its PBT. Thermal environments also show a greater variation due to elevation, as temperature drops approximately 1°F (0.5°C) for every 300 feet (91 m) of altitude, and more variation between night and day than in the lowlands; this requires changes in the reptile's behavior and physiology involving lowered optimum temperature for activity, increased basking time, earlier entry into hibernation, and later emergence.

Thermoregulation involves movement, and reptiles must continually move between niches in their territory to benefit from the varying conditions and temperatures. The study of captive snakes and lizards has shown that continued exposure to temperatures higher than the optimal can be very dangerous, but surprisingly it also proved that long exposure to preferred temperatures from which there was no escape was also unhealthy. The ideal arrangement for these creatures is brief periods in their optimal zone, followed by exposure to lower temperatures, but they must be free to choose.

The higher temperatures of the lowland tropics and subtropics favor a cold-blooded way of life, but although most reptiles do live in the world's warmer regions, the temperate zones are home to a large number of lizards, snakes, and turtles, and the temperate deserts are especially rich in species. The temperate zones are the two intermediate latitude regions of the earth, the North Temperate Zone which lies between the Arctic Circle and the Tropic of Cancer, and the South Temperate Zone between the Antarctic Circle and the Tropic of Capricorn. Their climate is defined by four seasons, often with a great variation between winter and summer. Animals hibernate in response to low temperatures, so it is naturally a form of behavior that has evolved in temperate regions where freezing is a feature of winter. Unable to migrate, when the temperature drops close to freezing all reptiles must hibernate or die, as their environment cannot provide the conditions conducive to activity and life—the heat to raise their metabolism and the availability of food.

The highest concentration of reptiles in the northern temperate zones occurs in the deserts, which are ideal for cold-blooded animals because of their high daytime temperatures and abundance of sunshine, but life in the desert requires strict regulation of body temperature and conservation of water, allowing reptiles to thermoregulate within their preferred range. Temperate deserts are very cold in winter and their reptiles must hibernate, but they also get extremely hot in mid-summer and there is a limit to the amount of heat reptiles can withstand. A body temperature of 104°F (40°C) is lethal to many lizards although a few desert species can withstand temperatures over 113°F (45°C). One very heat-resistant species is the desert iguana, which alternates between sun and shade to regulate its temperature, and is active when its body temperature is between 110° and 115°F (43.5° and 46.1°C), and its lethal temperature is 118°F (47.2°C).

The temperate-zone reptiles are by no means delicate creatures, unable to survive in regions experiencing long, cold winters. In fact in northern Eurasia the adder and the viviparous lizard range into the far north, even within the Arctic Circle, where it is barely warm enough for three months of reptilian activity each summer. The soil is so cold, being permafrost with just a shallow surface layer thawing in midsummer, that it is not a good incubation medium for eggs, so these reptiles incubate them inside their bodies and give birth to live young. Both species, and the wood frog in North America, are the northern-most cold-blooded land vertebrates, but the painted turtle has also evolved to withstand much lower temperatures than the lizards or snakes.

Adder Britain's only venomous snake, the adder has the most northerly distribution of all snakes occurring within the Arctic Circle in Siberia. The short and cool summers there restrict its active period to about 4 months yearly, and although primarily nocturnal it sun-bathes during the day to raise its temperature high enough to search for mice and voles.

Photo: Konstantine Kikvidze, Shutterstock.com

Cold-blooded animals convert practically all of their food into body mass, so the reptiles' metabolism allows them to quickly acquire a store of fat to provide energy for many months, whereas warm-blooded creatures use most of their food as energy to maintain their constant temperature. As metabolic energy is not used to heat the reptile's body its energy requirements are many times lower than that of the birds and mammals, and it has one-seventh the metabolic rate of similar-sized mammals, and therefore needs only one-seventh the amount of food. Reptiles are therefore less dependent upon a regular supply of food than mammals and can enter hibernation or estivation almost at will, surviving on a small amount of stored fat, which lasts longer as they need less energy to survive.

Most reptiles avoid freezing by finding a place where they are protected from the frost. Soil and water act as heat reservoirs and hibernators make full use of their protection. The aquatic turtles stay in the water beneath the ice, where the temperature is just above freezing, however cold it is at ground level. Terrestrial hibernators, such as the glass lizard, Gila monster, and desert tortoise, burrow into the soil below the frost line to benefit from the even temperature and humidity there. Others, like the garter snakes in their Manitoba limestone sink-holes, use natural crevices or caves, which give access deep into the earth. Many reptiles make use of the abandoned holes of rodents (especially ground squirrels) and of carnivores (badgers, foxes, and skunks) to avoid the freezing temperatures above ground.

A period of torpidity is now essential for the successful reproduction of most reptiles, even those living in the tropics where temperature variation between day and night may be only a few degrees. Reptile keepers are well aware that their snakes do not breed as well, or even at all, if they are not cooled for a short period at the right time of year. They call this brumation, a herpetological term for the artificial cooling of captive reptiles for short periods to simulate their natural seasons and thereby improve breeding success. It produces a state of semi-torpor without major physiological adjustment in which the animal is less active and does not eat or drink. Several reptiles can survive brief periods of freezing ambient temperatures, which may occur overnight. The freezing point of a reptile's body fluids is 31.1°F (-0.5°C), and extracellular ice then forms. Like other vertebrates it is unlikely that any reptile can survive the formation of intracellular ice, but the formation of ice crystals outside the cells during hibernation is known to be tolerated by several species and therefore quite possibly by others. Hatchling painted turtles can withstand the freezing of their extracellular fluids for several months, whereas red-sided garter snakes—the most widely distributed North American reptile—have survived after being frozen for a few hours, and box turtles for several days.

Although reptiles are not faced with the same degree of risk from desiccation in hot dry environments as the moist-skinned amphibians, and bask in the sun more frequently and for longer, most species must avoid intense heat and drought. This is not only necessary in tropical arid regions, but also in temperate deserts, which may get even hotter than the tropics in midsummer. They therefore estivate in response to extreme heat and dryness, and although it is mainly behavior common to the warmer regions of the world, reptiles in temperate deserts also become torpid to escape extreme summer heat. Also, like certain amphibians of the northern temperate regions some reptiles are forced to hibernate and estivate in the same year. A classic example of the behavior of temperate-zone reptiles in response to climate is provided by the cold-blooded inhabitants of the Craters of the Moon National Monument, 90 miles west of Idaho Falls, where lava eruptions have created a rugged landscape of lava tubes, cones, craters, and buttes. It is a fine example of a temperate arid region where the climatic extremes of winter and summer force the reptile population to hibernate and estivate. The midsummer temperature of the lava on which the terrestrial animals live can reach 170°F (76°C), and is worsened by 30 mph winds and scanty rainfall in June, July, and August. In winter it is just the opposite, with subfreezing temperatures and up to 24 inches (60 cm) of snow. The resident reptiles, which include gopher snakes, western rattlesnakes, horned lizards, western skinks, and sagebrush lizards, all hibernate to avoid the cold, and then about three months after their emergence in March are forced to estivate for almost three months.

Water conservation is extremely important to desert reptiles. They lack sweat glands and reabsorb moisture from their wastes and excrete semi-solid feces. Animals produce ammonia, which would be poisonous if allowed to accumulate in their bodies so it is converted into urea which is less toxic and can be stored longer. The terrestrial reptiles, which need to conserve water, excrete urea with their feces after converting it to semi-solids. Conserving water in this manner, plus their low metabolic rate and the ability to store the fat to fuel many months of hibernation and estivation, has allowed them to colonize deserts, where they are the dominant form of vertebrate life.

Reptile's lay shelled eggs that must be fertilized internally, unlike the amphibians whose eggs are fertilized by the male after being laid. The tortoises have penises, and the snakes and lizards have paired copulatory organs called hemipenes, although only one is used at a time. The fertilized egg develops in one of three ways. When the environment is suitable for external incubation—in sand warmed by the sun for example, or the fermentation heat inside a manure heap— some reptiles simply lay their eggs, and these are said to be oviparous. The cool substrate, and even permafrost, of the high north is not conducive to incubating eggs, so in the reptiles that live there, and many others farther south also, the eggs are "incubated" within the female's body and the young wriggle out of the egg's membrane as soon as they are laid. These reptiles are said to be ovoviviparous. In the third group the young are born alive after developing within the mother's body and being nourished by her in a manner similar to that of the placental mammals. They are said to be viviparous reptiles. Egg-laying in reptiles is considered to be a more primitive characteristic than giving birth to live young, as there is a greater risk of loss of the whole clutch from predation.

Most snakes and lizards lay eggs with parchment-like shells, whereas the chelonians lay either hard- or soft-shelled eggs depending upon the species. Some freshwater turtles can produce fertile eggs several years after being mated, having stored the sperm within their bodies. Another unusual feature of egg incubation in some species of snakes and lizards is that the incubation temperature determines the sex of the hatchlings, with temperatures above normal producing female offspring, and males when they are incubated at below the optimal temperature.

■ LIZARDS

Lizards are considered the most successful of the modern reptiles, and have a wide distribution on all the continents except Antarctica, and on many oceanic islands which they reached accidentally on driftwood. They have evolved to survive in a variety of habitats, including rain forests, grasslands, deserts, and swamps. There are arboreal lizards, terrestrial ones, some that are almost aquatic, and others that spend most of their lives underground and are rarely seen. Their greatest concentrations are in the equatorial regions, where they occur from sea level up to 15,750 feet (4,800 m) on the mountainsides.

Lizards have a wider north–south distribution than any other reptiles, and have also colonized temperate zones where their ability to hibernate allows them to live in regions with short summers and long, harsh winters. They are especially plentiful in semiarid regions with very hot summers, such as the southwestern United States, where they estivate during the hottest periods, but they vary in their ability to withstand extremes of temperature. The desert iguana can tolerate a body temperature of 115°F (46.1°C) and monitor lizards in Australia's Gibson Desert can survive core temperatures of 117°F (45.4°C). At the other extreme the viviparous lizard lives within the Arctic Circle in Eurasia, and the northern prairie skink survives in southern Manitoba where winter lows can reach −40°F (−40°C).

The lizards lack both outer ears and external ears. Their sense of hearing, at least of airborne vibrations, begins at the tympanic membrane or eardrum, which is visible on the surface of the head, and covers the middle ear cavity. Airborne vibrations are detected by the membrane, while ground vibrations are picked up by the small quadrate bone (which joins the upper and lower jaws); both of these in turn vibrate the tiny bones of the middle ear that pass the information to the cochlea in the inner ear, which then transmits it to the brain along the auditory nerves. Geckos' ears are less visible, but can be seen as small holes on the sides of the head, and they have well-developed middle and inner ears and good hearing. In association with their hearing sensitivity geckos are noisy creatures, with vocal chords unlike other lizards, and with loud voices for their size.

Lizards have scaly skins, which are hinged and flexible, like the snakes. Their scales, composed of thickened epidermis, may be very smooth and arranged in rows as in the snakes, or can be ridged, spiny, or keeled and irregular. The manner in which they shed or slough their skins also varies, and in most lizards it is shed in pieces or patches and may be eaten. In some species, such as the horned lizards, it is an almost continual process with single scales being shed regularly; while in others it happens every few weeks, and is often assisted by the animal itself pulling at the skin or inflating its body to stretch and loosen it. A few lizards shed their skins whole several times each year.

The presence of limbs generally distinguishes lizards from snakes, for most species have legs, usually strong well-developed ones; but there are also many legless forms in which their vestigial legs are not obvious externally and they therefore resemble snakes. However, these animals can be identified as lizards by the presence of closable eyelids, which of course snakes lack. Some lizards have transparent fused eyelids like the snakes, but most have movable ones, in a few species with a transparent bottom lid so they can still see with their eyes closed, and some of the burrowing species are blind.

Many species have a uniquely lizard habit called autotomy, which is the ability to regrow a tail lost after being seized by a predator, which is a valuable escape asset. They have a thin band of unossified tissue between each tail vertebra which acts as a break point when the tail is held. The day geckos can regenerate lost tails very quickly, often in three weeks, but the regrown tail never perfectly matches the original.

The lizards of the temperate regions that experience subfreezing winters cannot possibly survive without hibernating. Torpor involves storing food internally and finding suitable hibernacula where they will be protected from freezing. Unlike at least one other reptile—the painted turtle—and several amphibians that can synthesize cryoprotectants or anti-freeze compounds (allowing them to survive many degrees of frost), it is unknown if any lizards possess this ability, although the viviparous lizard may as it lives within the Arctic Circle. Many species have evolved to cope in temperate climates—to grow, reproduce, and find sufficient food during a short summer to store fat for the energy needed for a long winter. In all except the most southerly regions of North America lizards must hibernate[5] to survive, although the length of their dormancy varies according to the local climate.

In the northern regions of Eurasia the resident lizards must similarly hibernate all winter, with local conditions also dictating the length of their torpor; the record belongs to the viviparous lizard, which hibernates for nine months at the northern edge of its range. It is absent from the Mediterranean area, but occurs at higher elevations in the more southerly latitudes of its range, up to 11,500 feet (3,500 m) in the Alps, where it hibernates for at least six months. Reptiles appear generally less adapted for activity in very low temperatures, unlike many amphibians, although map turtles (*Graptemys geographica*) are very slow to become torpid and have been seen moving around beneath the ice in northern lakes.

Despite the ability of some arid-land lizards to withstand exceedingly high core temperatures, many species estivate to escape the extreme heat of desert regions. Skinks and geckos store fat in their tails for estivation as well as hibernation, and the Gila monster gorges after emerging from hibernation in April or May, stores fat in its tail, and estivates until the late summer rains arrive. In Brazil the teguexin or tegu (*Tupinambis teguexin*) estivates during the cooler, dry season, becoming dormant and hiding in its burrows when the temperature drops below 77°F (25°C).

Some of the Species

Geckos

The family *Gekkonidae* is a very large one, containing about 700 species of which three-quarters are nocturnal. They are mostly small, soft-skinned lizards, which range in size from some of the tiniest lizard species less than 2 inches (5 cm) long to 16 inches (40 cm) in total body length. The structure of their feet allows them to scale smooth surfaces. Most have sharp claws, and the undersides of their toes have pads covered with microscopic filaments called setae, which resemble stiff hairs. Interaction between these and the surface create an intermolecular force allowing them to walk up vertical glass. However, geckos' feet vary according to their habits and not all of them have glass-climbing toes. They depend upon sight to hunt at night and most species have large and prominent eyes, with large pupils, lens apertures, and corneas, which improves their ability to gather light although their visual acuity is reduced.

Geckos are all insectivorous or carnivorous and are stimulated by the movement of their prey. They lack the snake's constantly flicking forked tongue, and although they have vomeronasal organs these are probably not as useful to them as they are to the diurnal lizards and the snakes where the tongue transports scent particles to them. These organs are more concerned with tasting the gecko's food to determine palatability. Contrary to the general rule that lizards have eyelids, those of most geckos have fused and are immovable, remaining as a clear membrane which they clean with their tongues. The members of the subfamily *Eublepharinae*, however, have eyelids and are often called eyelid geckos.

Geckos are mainly lizards of the tropics and subtropics, but a few live in temperate regions that experience very hot summers and cold winters, where they must hibernate and then possibly estivate for short periods in midsummer, although their nocturnal habits generally preclude this. In the United States native geckos are restricted to the Southwest, from Texas west into southern California, plus southern Florida and the Keys. Like the alien anoles there, the southeastern states are also home to several exotic species of geckos including the Indo-Pacific gecko (*Hemidactylus garnoti*) and the Mediterranean gecko (*Hemidactylus turcicus*), which are established there. In Europe geckos are found mainly in the milder regions bordering the Mediterranean, where they do not need to hibernate.

Leopard Gecko (*Eublepharis macularis*)

The leopard gecko is a terrestrial lizard of the sandy and rocky deserts and arid grasslands of southern Asia, including Afghanistan, Pakistan, and western India, where it escapes the intense summer heat by being nocturnal and crepuscular, and spends the hot daylight hours hidden in a burrow or beneath the rocks. It stores fat in its tail and estivates during the hottest part of the summer, then hibernates during the short winter. It reaches a maximum length of 10 inches (25 cm), and is tan-colored, the adults being liberally covered with small dark brown spots, while the juveniles are usually banded with dark brown. This gecko has a long association with man, being eaten throughout the Far East in the belief that it will cure asthma and other lung conditions; more recently in North America it has become one of the most popular pet lizards due to its ease of maintenance and reproduction. After two decades of selective breeding, leopard geckos are now available in a number of new pattern and color mutations including tangerine hypomelanistic—in

Leopard Gecko *The leopard gecko's home in the rocky deserts of Afghanistan and Pakistan is a very harsh environment, subjected to intense heat in summer and bitter cold in winter. It is forced to avoid the extremes and estivates and hibernates, relying on the fat stored in its tail for these dormant periods.*
Photo: Barbara Brands, Shutterstock.com

which the black pigment is reduced—and leucistic, which is a white color variant. Leopard geckos belong to the subfamily *Eublepharinae*, the members of which have moveable eyelids and visible outer ears, but they lack the filaments on their toes and cannot walk up smooth vertical surfaces.

Desert Banded Gecko (*Coleonyx v. variegatus*)

This gecko is a race of the western banded gecko which is widespread in the desert regions of the southwestern United States and neighboring Mexico. Its actual range is the Mojave and Sonoran deserts, where it lives in rocky or sandy terrain, semiarid scrub, and oak and pinyon woodland to an elevation of about 5,000 feet (1,525 m). Like most geckos it is nocturnal, and has vertical pupils. The desert banded gecko is about 6 inches (15 cm) long and has a translucent, pale-pink skin with brown bands when young; but these change into spots and blotches with age.

It hides under rocks and debris by day, or in moist rock crevices or rodent burrows, appearing at dusk to hunt for beetles, spiders, flies, young scorpions, crickets, and grasshoppers. Prey is captured after a stalk and a lunge, not the flicking out of a sticky tongue, but it does wipe its face with its tongue after eating. It also eats its dead skin when this sloughs off in patches in the normal lizard fashion. In good times these geckos store fat in their tails, which become quite extended, and this sustains their lowered metabolism during torpidity that may last for nine months each year. They hibernate from November to early April, and then estivate in midsummer, mating and laying just two eggs, which is a gecko characteristic, between emerging from hibernation and entering estivation.

Iguanids

These lizards are the members of the family *Iguanidae*—the iguanas and their relatives—a group of mostly New World species with a few representatives in Madagascar and Fiji. Most North American lizards belong to this large family, but they are mainly of central and southerly distribution, especially in the Southwest, with just a few species extending north almost to the Canadian border and one—the horned lizard—just crossing the border. Members of this family also occur throughout Central and South America south to Uruguay, plus the West Indies and the Galapagos Islands. Although they are better known by the name iguana, this is always associated with the large green "pet" lizards, and only one of the many North American members of the family is actually called iguana, namely, the desert iguana (*Dipsosaurus dorsalis*), so iguanids is a more appropriate name for the family members.

The many species of anoles also belong to the family *Iguanidae*. The green anole is the only native species in North America, occurring from Florida north to North Carolina and Tennessee, and is not known to hibernate, not even in Tennessee. It remains active all winter, preferring south-facing rocks where it can bask in the morning sun, and then retires to a rock crevice at night. The other anoles of Florida such as the Cuban brown anole (*Anolis sagrei*) and the Jamaican giant anole (*Anolis garmani*) are all recent immigrants and have no history of hibernation in their homelands. The family has been reviewed and reorganized recently, and several of the long-accepted members, such as the horned toads, have been removed to other families by some zoologists.

Pygmy Horned Lizard (*Phrynosoma d. douglassi*)

The only iguanid occurring in Canada, this small lizard is one of several races of the short-horned lizard, which are also called horned toads. It has pointed scales with up to twelve projections or "horns" arising from the back of its head, and a single row of fringed scales on the sides of its body. It is well camouflaged, its coloring generally matching the substrate, as it can change color from light to dark to match its surroundings. To deter predators, it spurts blood from its eyes—actually from a channel at the base of the third eyelid. It can also inflate itself by gulping air and may try to "attack" an aggressor with its horns, by twisting its head from side to side.

The pygmy race of the horned lizard occurs in extreme southern British Columbia and from there south through the dry regions of Washington and Oregon to northern Nevada and California. It lives in semiarid plains and grasslands and on rocky hillsides, and is well camouflaged for its sandy and stony habitat, being pale brown, dark brown or dark bluish-gray above. It is a very cold-tolerant animal, and twisting from side to side forces itself down a few inches into the sandy soil, which is all that protects it from subfreezing temperatures. Hibernating from

Pygmy Horned Lizard *An unusual spiny reptile of the semi-arid lands of southern British Columbia south to California, the horned lizard hibernates from November to March in the north, burrowing just a few inches into the sandy soil for protection from the elements. As protection from predators it inflates itself and squirts blood from a channel at the base of its third eyelid.*
Photo: Steve Shoup, Shutterstock.com

November to mid-March at the northern end of its range, it is then active until June when it digs into the sand again and estivates for six weeks to escape the heat of midsummer. This phase ends when the slightly cooler temperatures and rains arrive in August. The horned toad is mainly an ant-eater, but it stops eating a few days before commencing hibernation, for digestion ceases at low body temperatures. Mature males become sexually active about a week after emerging from hibernation, which may last four months. Like the other iguanids, the base of the male's tail becomes swollen when he is in breeding condition.

Collared Lizard (*Crotaphytus collaris*)

This is a very sturdy lizard with a large head and powerful jaws, and together with its long tail it reaches a length of 14 inches (36 cm). Its most conspicuous feature is a black-and-white collar, and its background color varies, being either olive, light brown, yellowish, or even having a bluish or greenish tinge, with broad dark crossbands and spots over the body, and small spots on the face, legs, and tail. It is an arid land lizard which frequents the canyons, gullies, and rocky mountain slopes of the Great Basin south to Baja California and east to Missouri and Texas. It loves to bask and maintain a vigil from the tops of large boulders, usually in areas where sparse vegetation permits a good view. It is a fast and agile species that actively chases its prey (which is mainly invertebrates and small lizards), often running upright on its hind legs, but it also eats young mice and nestling birds. The collared lizard hibernates in the northern parts of its range—Idaho, Utah, and Colorado—from October to March, and for progressively shorter periods farther south. In many areas it may estivate in midsummer, especially in Death Valley where the temperature has reached 134°F (56°C) (see color insert).

Chuckwalla (*Sauromalus ater*)

The chuckwalla is a large, plump lizard with a flattened appearance and very rough sandpaper-like skin, with loose folds on its neck and sides. It has a thick, blunt tail, reaches a total length of 17 inches (43 cm), and weighs up to 2 pounds (0.9 kg). Its forelimbs and the front of its body are dark gray sometimes spotted with paler gray, and the rest of the body is light reddish-gray and the tail is pale yellow. The chuckwalla is adapted for very high temperatures, and lives in the flatlands and rocky hillsides of the Mohave and Sonoran deserts, from southern Nevada to northern Mexico. It is a diurnal lizard that loves to bask, which it does immediately upon appearing from its rock shelter in the morning, for it must raise its body temperature to the preferred range of 100°–105°F (38°–40.5°C) before its normal activities can begin. The most important of these is gathering food, and it is almost totally herbivorous, eating leaves and flowers, the foliage of the creosote bush being a favorite food, and occasional insects. When alarmed the chuckwalla wedges itself into a rocky crevice by inflating its body. The chuckwalla can tolerate high air temperatures and when at risk of its own body temperature rising above 105°F (40.5°C) it seeks shade under rocks and in crevices, but it does not estivate. The southwestern deserts experience frost in winter, however, and the chuckwalla hibernates—in a crevice or among rock piles from November to February.

Lacertids

These are a family of about 180 species of small, long-tailed lizards from Eurasia and some Pacific Islands, with a wide latitudinal range from within the Arctic Circle into the tropics. Most species have thin bodies and big heads with the males generally having larger heads and shorter bodies than the females. The lacertids are very fast runners and agile climbers, and their forked tongues have olfactory powers like the snakes, carrying scent particles to their Jacobsen's organs for identification. They are highly insectivorous and very territorial, and most species lay their eggs in sand. The majority of European lizards belong to this family, including the viviparous lizard, which is the world's most northerly distributed reptile.

Eyed or Ocellated Lizard (*Lacerta lepida*)

This is Europe's largest and most impressive lizard, with a head and body length of 10 inches (25 cm), and a tail almost twice as long. Green with black stippling and blue spots ("eyes") on the flanks, the males are larger and more brightly colored than the females; whereas juveniles are pale greenish-yellow with black-edged white spots. The eyed lizard is mainly a lowland animal of southwestern Europe and the coastlands of northwestern Africa, and in most of its range it frequents arid habitat, such as rocky scrubland, olive groves, and dry grassland. In northeastern Spain, however, it also occurs in cool and moist country and has been found at up to 6,560 feet (2,000 m) in the Pyrenees. It is diurnal and mainly a terrestrial hunter of small rodents, snails, and invertebrates, but it climbs trees to reach birds' nests, and during the summer months also eats berries. In the southern parts of its range the eyed lizard hibernates from November to late February, in abandoned rabbit burrows or in hollow trees or beneath logs, but in the cooler regions this torpid period begins at least two weeks earlier and extends until the end of March.

Green Lizard (*Lacerta viridis*)

The green lizard lives in central Europe from northern Spain to Holland and then across the continent to Turkey. It is the largest lizard north of the Alps, with males reaching a total length of 17 inches (43 cm); the tail may be 10 inches (25 cm) long. The male is a very attractive animal, clad in brilliant green, finely stippled with black, and has a blue throat, whereas the female is dull greenish-brown. Green lizards prefer a dry and sunny environment, and are at home on stony and wooded hillsides, scrubland, rocky river valleys, and vineyards, where they are active by day and take refuge at night in abandoned rodent or rabbit burrows or beneath rocks. Totally carnivorous, they eat invertebrates, young rodents, snails, and nestling birds. Like all European lizards which are forced to hibernate, courtship and breeding commences soon after emergence, and the female lays up to twenty eggs, which she buries in a pit in sandy soil. The adult lizards hibernate in September, but hatchlings of the year stay active until the end of October, this later entry into torpor allowing them to store more fat for the winter and improve their chances of survival; both adults and young emerge in April. This species has been introduced and is now established in the vicinity of Topeka, Kansas.

Common Wall Lizard (*Podarcis muralis*)

This is one of the common lizards of mainland Europe, occurring throughout the continent from Spain east to Romania, but excluding Poland and Scandinavia, and it is now established in the vicinity of Cincinnati, Ohio. It is a thin, brownish-gray lizard, just 9 inches (23 cm) long including its tail, and has very variable markings. Most individuals have rows of dark and yellowish-white spots along the length of the body and base of the tail, but in some regions they may have blue spots on their shoulders, and in other areas the spots on their backs and sides are often bright pale green. The belly is usually pinkish-red.

The wall lizard is a gregarious species, usually found in small colonies in suitable habitat, which includes dry rocky and stony country, ruins, and dry stone walls in which the slabs or rocks are not cemented together, leaving many hiding places suitable for small lizards. It is totally diurnal and sleeps at night in the crevices, where it also hibernates. In the northern parts of its range this commences in late October and lasts until the end of March, but in sunny locations it has emerged in February to bask on the sun-warmed rocks. It is a very hardy and adaptable species if suitable hibernation sites are available and there is adequate food to provide winter fat reserves. The introduced colony in Ohio has survived for half a century despite several winters of record-breaking severity. The wall lizard lays its eggs under stones or in rock crevices, or buries them in sun-warmed sandy soil. It is mainly insectivorous, but also eats snails, baby mice, and nestling birds, and is highly cannibalistic of newly hatched conspecifics.

Viviparous Lizard (*Lacerta (Zooteca) viviparia*)

This is the most common lizard in northern Europe, the only reptile found in Ireland, but is unusual for two reasons—its choice of habitat and the fact that it is the world's northern-most reptile. It ranges from northern Scandinavia and Russia (from latitude 70°N, about 200 miles within the Arctic Circle), south to Spain and Yugoslavia, and eastward to northwestern Asia. It has also been found at up to 11,500 feet (3,500 m) in the Alps. Unlike the other European lizards it prefers a moist habitat including moorland, peatbogs, and marshes, where it enters water frequently and is a good swimmer. It is a coarse-scaled, long-bodied and short-legged lizard, just 7 inches (18 cm) long, and when mature has dark brownish-gray coloring with alternating longitudinal rows of buff spots and small black markings, these often being tipped with pale yellow. Males have an orange belly with dark spots, and the females have creamy-white bellies.

The viviparous lizard's diet includes worms, flies, and beetles and their tiny young eat aphids, but they do not become active until their body temperature reaches 86°F (30°C), so the high northern distribution creates thermoregulation problems for them. The short summers force them to hibernate for almost nine months each year, allowing little time for the breeding cycle, and certainly no opportunities for eggs to hatch in the cold soil. Consequently, the young lizards—up

Viviparous Lizard *The northern-most reptile, this lizard lives within the Arctic Circle in Europe and Russia, where it is forced to hibernate for almost 9 months each year to avoid low temperatures. As the soil is too cold to incubate buried eggs in the normal lizard manner, the female retains them within her body, where they complete their development and hatch immediately when they are laid.*
Photo: Steve McWilliam, Shutterstock.com

to eight per clutch—develop internally within the eggs and usually break out of the membrane as soon as the eggs are laid. They must then quickly store fat before lowered temperatures force them to hibernate. Their period of dormancy varies according to latitude and altitude, and in contrast to those resident in the far north, Spanish populations are torpid for only three months, whereas at high elevations in the Alps they sleep for at least seven months. Hibernation usually occurs under a layer of leaves and the insulating snow, but they may enter holes and crevices.

Glass Lizards

This small family of lizards includes some with tiny limbs and others that are completely legless. They are distributed throughout the world, although mostly in the tropics, and the few northern species must hibernate to survive the winter. With their elongated bodies the glass lizards resemble snakes, but have typical lizard heads with eyelids and ear openings. In keeping with their shape their locomotion is fast and snake-like, and like the snakes they have forked tongues and are all carnivorous. Their scales are reinforced with small body plates called osteoderms, and the result is a stiffening of the body, so glass lizards feel firm and rigid when handled, not supple like snakes. To compensate for this rigidity they have a flexible groove covered with granular scales along the sides of their bodies, allowing expansion when breathing and distension for large meals and when they are nurturing a clutch of eggs. Glass lizards have very long tails, that may be double the length of their bodies, but they are brittle and break off easily, so it is rare to see a full-length adult. The tails can be regenerated, although they are never as long as the original and have a pointed tip; the lizards are consequently often called "horn-snakes." They are burrowers and their diet consists of snails, insects, other small lizards, snakes, and birds' eggs and nestlings.

European Glass Snake or Scheltopusik (*Ophisaurus apodus*)

A heavy-bodied, diurnal snake-like lizard, from southeastern Europe and ranging eastward through Turkey and Syria to central Asia, the glass snake grows to almost 4 feet (1.2 m) long. It has a very firm, olive-green or brown body and lacks the snake's suppleness due to its tight-fitting scales. To compensate for this it has a lateral fold of soft skin along the sides of the body (but not the tail) which can expand for breathing and eating. It has tiny vestiges of hindlimbs as two spurs near the vent, uniform lizard-like scales, an unmistakable lizard's head with nonexpandable jaws, and prominent ear openings. A burrower that hides in the soil and is seldom seen, it is a very aggressive carnivore with a powerful bite, which tackles anything smaller than itself; its diet includes invertebrates, snails, nestling birds, eggs, small mammals, and other lizards and snakes. It has a long and very brittle tail (hence its name), which forms at least half of its total length, but is shed easily and regrows, although not to its original length. Glass snakes lay eggs, usually no more than six at a time, under a rock or in a pile of rotting vegetation, and the mother guards them for their thirty-day incubation period. It is an animal of dry habitat, including arid stony steppes, scrubland, and dry forest, and is crepuscular or nocturnal, hiding by day under debris, leaves, and logs. The glass snake hibernates throughout its range, and in Turkey its dormant period lasts from early November to mid-March, in the security of crevices, rotting tree root systems, or rodent holes.

Western Slender Glass Lizard (*Ophisaurus a.attentuatus*)

This lizard occurs from Illinois south to the Gulf Coast and west to the Rockies. It prefers dry country, and especially sandy soils into which it can easily burrow, and is found in grasslands, open oak woodlands, and sandy prairies. It is a large but slender animal, reaching a length of 42 inches (1.07 m), of which the head and body length is only one-third, the remainder being tail, although there is no clear demarcation between them. Its body color is variable and can range from tan to dark bronzy-brown. There is a thin brown stripe running the length of the spine, and two wider dark stripes along the sides of the body and tail, and sometimes finer stripes on either side of the lateral groove; the belly is pale yellow. This species hibernates from October to early March in the north, burrowing into loose soil to escape the frosts. Its tail is very brittle and breaks off easily, which can happen when it shakes vigorously if it is held. Consequently, most adults have short, regrown tails. This glass lizard is primarily insectivorous, preferring grasshoppers, crickets, and beetles and their larvae, but may also eat nestling birds and young mice.

Slowworm (*Anguis fragilis*)

The slowworm is not a worm, but a legless lizard that resembles a snake, with a thin, smooth and shiny body and an indistinct head. It reaches a length of 21 inches (53 cm) and like the glass lizards can shed its tail, and it has eyelids. Males vary in color from gray to bronzy-brown, and have a pale belly, whereas females are browner and have dark underparts, but both may have blue spots on their bodies. It is most active at dusk and dawn, when it searches the leaf litter and loose soil for soft-bodied prey such as worms and slugs. The slowworm has a wide range, occurring in Europe, southwestern Asia, and north Africa. It is nocturnal and hides by day under rocks, leaves, or logs, although it may appear briefly to sunbathe. In northern Europe it hibernates from October to early March, individually or communally and sometimes together with snakes, in crevices, holes, and under piles of leaves. Although mating occurs soon after emergence from hibernation, the eggs are not fertilized until June when they enter the oviduct. The slowworm has up to twenty live young at a time, which develop inside the eggs within the mother's body and break out of their membranous covering as the eggs are being laid.

Skinks

Skinks are characterized by smooth and shiny bodies and small heads, and appear neckless as their necks are the same thickness as their bodies. They are all alert, active, and very quick in their movements. Most have tiny limbs, but several burrowing skinks have lost their legs. Their scales are like the glass lizard's—reinforced with bony plates or osteoderms. Like the snakes their tongue is forked and flicks out constantly gathering airborne chemical scents. They are terrestrial and hide under rocks or logs at night. The females of several species guard or "brood" their eggs for the whole incubation period, which is very unusual behavior for reptiles. The North American skinks lack eyelids, but some skinks have a transparent window in their bottom eyelid and can see with their eyes closed.

The family *Scincidae* is the largest group of lizards with about 1,200 species, most of which are diurnal. They are mainly tropical and subtropical animals but a few live in temperate regions where they must avoid the cold. In North America there are nineteen species of skinks, most of which hibernate in winter, the longest period of torpor being undertaken by the five-lined skink (*Eumeces inexpectatus*) and northern prairie skink (*Euemeces septemtrionalis*) both of which range north into southern Canada, where they must avoid six months of winter. In regions with extremely hot summers such as Death Valley and the central Great Basin they must also estivate during midsummer. Similarly, the Great Plains skink (*E. obsoletus*) and the western skink (*E. skiltonianus*) hibernate and then not long afterward estivate. In Europe, skinks are mainly concentrated around the Mediterranean Sea, but some range northward into central Europe and must hibernate. The snake-eyed skink (*Cryptoblepharus poecilopleurus*) of Slovakia, Serbia, Hungary, and Romania is one of these, hiding under fallen leaves and becoming torpid from October to April.

Australia is especially rich in skinks with over 350 species, but due to its relatively warm climate hibernation is rare. However, at least two species become torpid in cold weather, although they apparently have a greater-than-normal lizard tolerance for low temperatures. They both live in the high country of southern New South Wales and neighboring Victoria, where Coventry's skink (*Leiolopisma coventryi*) has been found dormant under leaves or in rotten tree trunks in midwinter; and the alpine water skink (*Eulamprus kosciuskoi*) also hibernates, but has been seen out on the snow on the slopes of Mt. Kosciusko at 3,000 feet (915 m). Australia's Red Centre is very hot in midsummer but it seems likely that the lizards living there change their habits to nocturnal activity to escape the heat, rather than becoming torpid for long periods. The normally diurnal Centralian blue-tongue skink (*Tiliqua multifasciata*) and Mueller's skink (*Lerista muelleri*) both become nocturnal for the summer.

Northern Prairie Skink (*Eumeces s. septemtrionalis*)

This race or subspecies of the prairie skink (the southern prairie skink being the other one) ranges from Minnesota and Wisconsin south to Kansas, but there is an isolated population in Canada—in Manitoba's Spruce Woods Provincial Park. It prefers a semiarid environment with sandy or gravelly soil, glacial-lake deposits, and outwashes. A secretive lizard, it digs down into the soil, or hides in the burrows of ground squirrels, and is rarely seen. Only 8 inches (20 cm) long, of which 5 inches (13 cm) is tail, it is olive to olive-brown with many dark and light stripes running from the head onto the base of the tail. In southern Manitoba, which can experience temperatures down to −40°F (−40°C) in January and February, the skink hibernates in rodent burrows or any crevice where it can get below the frost line, and is dormant from October to April. When it emerges in spring it must bask immediately to raise its body temperature to the level required for hunting invertebrates. It is diurnal but during the hottest part of the day hides beneath rocks, logs, or piles of debris. Its reproductive behavior includes guarding its eggs and then eating those that do not hatch.

Western Skink (*Eumeces skiltonianus*)

This skink lives in semiarid country from southern interior British Columbia south to Baja California, and from the Pacific Coast inland to Missouri. It has been found at up to 8,000 feet (2,440 m) on the north rim of the Grand Canyon. It is a tiny animal, just 3 inches (8 cm) long, with pale-gray underparts and sides, and is striped along the length of its back with narrow dark-brown and pinkish-buff lines, and a wider brown stripe along the spine. Its tail is buffy-brown in adults, tipped with bluish-gray, whereas hatchlings have a bright-blue tail. The western skink lives in dry grassland and rocky wooded hillsides and occasionally in moister areas. It is diurnal and terrestrial, and hides under rocks and logs when alarmed, and at night. It hibernates from October to March in the northern parts of its range, usually in piles of debris, in abandoned pack rat nests, or inside decaying logs. In some areas, such as Craters of the Moon National Monument and Death Valley, where on forty-three consecutive days in the summer of 1917 the air temperature was over 120°F (48.8°C), it is not surprising that it also estivates. It has no need to hibernate in Baja California, but may be inactive in midsummer.

Broad-headed Skink (*Euemeces laticeps*)

A common lizard of the central and eastern United States, the broad-headed skink can be found in woodland from Pennsylvania and Kansas south to Texas and Florida. It is the second largest of the North American skinks, just slightly smaller then the Great Plains skink, reaching a maximum length of 12 inches (30 cm) about half of which is tail. It prefers old-growth forest with lots of dead trees, and secondary growth where the logged stumps are well decayed; it is quite arboreal, often going high into trees in search of insects, the only native skink to do this. It is a very aggressive lizard, and males fight viciously in defense of their territory. Parental care is not a highly developed reptilian characteristic, except in the crocodilians, but like several species of skinks the broad-headed skink does guard its eggs until they hatch.

After laying from eight to twelve eggs in spring on a bed of decaying wood in a protected place such as a tree cavity, the female then remains close and guards them for their eight weeks' incubation. The hatchlings have a bold striped pattern and a brilliant blue tail, but the stripes and color fade with age, although females may retain a few stripes, and the adults are plain-colored in olive-tan, the males having very large, bright-red or orange heads and "swollen" cheeks. The broad-headed skink eats insects, spiders, young rodents, and small lizards and is apparently very fond of the larvae of the paper wasp; it is impervious to the wasp's sting, which cannot penetrate its hard scales. At the northern end of its range it hibernates from late October to March in tree cavities and in the rotting wood of old tree stumps.

Tuataras

The tuatara is a unique lizard-like reptile from New Zealand, a "living fossil" with a primitive body structure that is believed to have changed little in 200 million years. It differs from the lizards in having no visible ear openings, and in its unique teeth, having a single row in the lower jaw that fits into a double row in the upper jaw, and which are made of bone and attached to the outer surface of the jawbone. When its teeth are worn down, the tuatara still chews its food with its jawbones. Tuataras range in color from olive-green to brown, and have a scaly but soft skin with a crest of spines along the back and over the head. Unlike the snakes and most lizards they have a thick, un-forked tongue. Males grow to 24 inches (61 cm) long and weigh just over 1 pound (454 g); the females are slightly smaller.

Tuataras evolved in a temperate region that does not experience very hot summers. Their preferred range is 60°–70° F (15.5°–21° C) and temperatures over 80° F (26° C) have proved fatal. Therefore, even in such a mild environment they must avoid extremes and so are nocturnal and spend the daylight hours in petrel[6] burrows, and then they hibernate from June to August to avoid low winter temperatures. They store fat in the base of their tails to power their reduced metabolism, and the lifespan of captive animals has been shortened when they were not allowed to hibernate, probably because of obesity, as they continued to eat when they should have been dormant and using up their fat reserves.

The tuatara has a parietal or "third" eye, which is a photosensitive organ connected by nerves to the pineal body—a small endocrine gland in the brain—which has several functions. It acts as a light sensor, probably controls the animal's circadian rhythms, and is active in thermoregulation or temperature control and in triggering hormonal production, especially that which is concerned with reproduction. The eye is situated on top of the tuatara's head and is quite visible in young individuals up to the age of five months, after which it is covered with skin. It does not provide vision in the normal sense, as it has only a rudimentary retina and lens, but is sensitive to changes in light. Tuataras are very long-lived, reaching the age of sixty years, probably on account of their low metabolism, and also probably another example of long-term dormancy increasing an animal's life span.

Northern Tuatara (*Sphenodon punctatus*)

This is the most plentiful of the two species of tuatara, occurring on islands in Marlborough Sound, especially Stephen Island, and on islands in Hauraki Gulf, off the Coromandel Peninsula and in the Bay of Plenty. It is olive-brown with pale underparts. Although it may appear during the day to bask in the sun, it is quite nocturnal and rarely moves far from its burrow entrance when it emerges at dusk to feed. Its main source of food is the eggs and young of the shearwaters whose burrows it occupies, and the invertebrates that live in the soil enriched by the shearwaters' dung, and on the plant life which this supports. A favorite prey is the large-bodied, grasshopper-like weta, and it is also cannibalistic and eats its own young.

Heloderms

The Gila monster and the very similar beaded lizard are the world's only venomous lizards.[7] They inject their venom in a manner similar to the rear-fanged

Northern Tuatara *A primitive lizard-like reptile, the tuatara avoids extreme temperatures even in New Zealand's mild North Island by being active at night and hibernating during the Austral winter (June to August). It has a third "eye," a photosensitive organ on top of its head which is involved with thermoregulation and hormone production.*
Photo: Courtesy Department of Conservation, New Zealand, photo by Mike Aviss

snakes, with slightly enlarged teeth that are grooved to aid the flow of venom from the duct, and they must chew on their victim to force it into the bite wounds. Their bite is extremely painful but the neurotoxic venom from their paired and modified salivary glands is rarely fatal to healthy humans. These lizards are known as helo-derms after their family name *Helodermatidae*, and are the only survivors of a much larger group of venomous creatures that began evolving in North America's southwestern deserts about 40 million years ago, when the region was much wetter and covered with deciduous forest. As the climate dried they became fossorial to avoid the extremes.

They are stocky animals, with rounded heads, plump tails, and short but powerful legs. Their skin is granular or finely beaded, and their bright coloring warns potential predators to stay clear. The heloderms are solitary lizards, crepuscular and nocturnal in their activities, that live in desert and semiarid regions, usually where there is scrub cover. Their diet consists of invertebrates, lizards, snakes, young rodents, birds' eggs, and nestlings. They are voracious animals that can eat up to one-third of their body weight at a time, and therefore soon lay down a store of fat for their long periods of torpor. Living a mostly subterranean life, they appear on the surface at night for a few months each year and then retire and survive on their stored fat. With such a short active life and a low metabolic rate when resting, they are long-lived and have reached the age of twenty years. The heloderms are egg-layers; they mate in the spring and produce up to twelve eggs in midsummer which they bury several inches deep in the sand. The incubation period is five months, but they sometimes overwinter and hatch the following spring. They have forked tongues, like the snakes and many diurnal lizards, and pick up scent particles which are transferred to their vomeronasal sensory organs and thence to the brain for analysis. Possibly they rely more upon their chemical sensing powers than on sight, except when hunting in daylight, which they do occasionally. Although primarily nocturnal, they actually have good daylight vision and a keen sense of hearing.

Gila Monster (*Heloderma suspectum*)

The Gila monster is the only venomous lizard in the United States, occurring in the Southwest and in neighboring northwestern Mexico, specifically the Mohave, Chihuahua, Sonoran, and Colorado deserts. It is mottled or banded with black, pink, and orange, and its tail has five bands. It reaches a length of 24 inches (61 cm) and may weigh 5 pounds (2.2 kg). There are two races of this species. The southern one lives in the Sonoran and Chihuahua deserts and is called the reticulate Gila monster (*H. s. suspectum*) in which the adults are more mottled and blotched, with a black face and a background color dominated by black. The northern race, the banded Gila monster (*H. s. cinctum*), lives in the Mohave Desert, and is distinctly black-banded on a light-pink background, but also has a black face. The Gila monster is probably the most successful of all reptiles at completely avoiding climatic extremes. It emerges from hibernation in April, when food is most plentiful, especially birds' eggs and nestlings. After gorging and storing fat very quickly it becomes torpid until the summer rains begin in July when it is active again at night, and mates and lays its eggs. It then hibernates in November and is not seen again until spring. When these lizards sunbathe they may only expose their head at the entrance of their burrow, as this warms up quickly due to its small body mass and the fact that it is aided by a constriction of the jugular vein until the head is very warm, when blood is then allowed to flow into the body. This safety feature allows the lizard to warm up while still in partial concealment from predators (see color insert).

Beaded Lizard (*Helodera horridum*)

This lizard is larger and darker than the Gila monster, quite dark in fact, with some specimens being almost completely black, but generally with bands of yellow spots around the body. It lives in the deserts and canyon country of western Mexico, especially in the arid regions along the Pacific Coast. It grows to a length of 3 feet (90 cm) and reaches a weight of 6 pounds (2.7 kg).

■ SNAKES

Snakes belong to the suborder of reptiles known as *Serpentes*, and with their ancestors the lizards form the order *Squamata*. There are almost 3,000 species, and they are wide-ranging animals, occurring on all continents except Antarctica, and extending north into the Arctic Circle in northern Eurasia, and south almost to the tip of South America. Yet they are absent from many islands on which lizards still occur, including Ireland and New Zealand. They evolved from ancestors with limbs, but exactly how, where, or even when this occurred is unclear. It was certainly over 100 million years ago that the first known snake-like animals appeared, and it may have been on land, since like the modern legless lizards, the loss of limbs and external ears and the characteristics of their skulls suggest a fossorial (burrowing) way of life. Or it may have occurred in water, the snakes' long necks, elongated bodies, and in some species a laterally compressed shape suggesting an aquatic orgin.

Whatever their origins, however, snakes are certainly unmistakable, with their elongated, scaly[8] bodies and sinuous movement, forked tongues, lack of external ears, and staring lidless eyes—as their eyelids are fused to form a transparent covering over the eye. Their bodies are supported by up to 300 vertebrae and associated ribs, which extend around the body and are attached to the ventral scales, their locomotion resulting from the body being literally pulled over the scales by muscular action. While there are legless lizards that resemble snakes, no snakes have legs, and many less obvious facts are also commonly known, such as their ability to swallow prey several times wider than their own girth and the highly lethal nature of many species. Less well-known features include their highly sophisticated senses of smell and taste, and heat-sensing; their lack of a breastbone, which allows the ribs to expand to accommodate large prey; the fact that most have only one developed lung—the right one—and the ability to shed their skin whole, a process known as sloughing or ecdysis. Surprisingly, snakes do have tails, which is the end section of their body beyond the cloaca. Their long, forked tongues, which are flicked out constantly through a groove in middle of the lower lip even when the mouth is closed, are sensory organs that carry scent particles back to the vomeronasal organs.

Snakes have pointed and recurved teeth in both jaws, which gives them a good grip on their prey, and they shed and regrow their teeth periodically, including the fangs of venomous species. Except in the rear-fanged snakes that "chew" on their prey to force venom into the wound, they do not chew their food in the accepted sense and are adapted for swallowing it whole, the flexible ligaments between the jawbones permitting a very wide gape and allowing them to swallow prey with a far greater circumference than their own heads. There are snakes that give birth to live young, and snakes that lay eggs, in some cases coiling around them and raising their temperature slightly through muscle contractions—the closest thing to parental incubation among the reptiles—but parental care of the young is not a snake trait.

While tropical lowland snakes are not adapted for hibernation and a major drop in temperature is harmful and potentially fatal, northern reptiles have evolved to hibernate when the ambient temperature drops close to freezing, and in fact must do so to survive. All northern snakes therefore hibernate if their environment experiences low temperatures. This certainly means the subfreezing temperatures such as those experienced in southern Canada and the northern United States, but it can also include snakes living in the milder winters of Southern California, such as the rattlesnakes accustomed to very high summer temperatures in the vicinity of Palm Springs, that hibernate although winters are quite mild. Snakes have only been able to colonize the seasonally frigid regions of the world, due to their ability to become torpid. If they cannot, then they will freeze and die. Only one species—the garter snake—is currently known to survive some freezing of its body fluids for short periods, although they are not as freeze tolerant as some northern frogs and hatchling painted turtles. It is quite likely, however, that further studies will reveal that several other snakes of northern latitudes are just as freeze tolerant. To avoid freezing temperatures snakes head for their hibernation dens or hibernacula as soon as temperatures begin to drop in the early fall. Their dens, which may be traditional sites used for many years, include limestone sink-holes, deep rock fissures, rodent burrows, hollow trees, rotting tree stumps, wood ant mounds, crevices in old buildings or bridge foundations, in fact virtually anywhere they are protected from the frost. Hibernacula are often used by generations of snakes, which mass there in large numbers, sometimes in the company of other species. On the northern prairies snakes such as the prairie rattler, garter snake, and hognose snake may hibernate for seven months each year, with almost another month spent in the vicinity of the den each day after their emergence in spring, so they can shelter there again at night. When they return to their dens in the fall, they are assisted by their Jacobsen's organs—their highly sophisticated sensory hunting mechanisms—which also have a social function in aiding hibernating snakes to form their groups and to return to the right den.

Two races of the common garter snake (*Thamnophis sirtalis*) live farther north than any other reptile in the Western Hemisphere, the eastern garter snake (*T. s. sirtalis*) in central Quebec, and the red-sided garter snake (*T. sirtalis parietalis*) in the Northwest Territories. In Eurasia the range of the adder or viper extends even farther north—to within the Arctic Circle. The garter snake is the most studied hibernating snake, and it has been determined that the temperature in its hibernacula rarely drops below freezing, but if it does, it is able to withstand a few degrees of frost, down to 29.3°F (−1.5°C) for a day or two, and it has survived the freezing of 50 percent of its body after being exposed to temperatures of 27.5°F (−2.5°C) for several hours. If the ambient temperature drops lower, garter snakes must move to a warmer part of the hibernacula or die. This abilty to withstand short periods of freezing is valuable to the garter snakes when they have left the hibernacula in spring, when nights can still get quite cool.

Despite their general need to hibernate when the temperature drops, there is still considerable variation among species of snakes regarding their optimum or preferred body temperature. For example, rattlesnakes in California's San Bernadino Mountains hibernate when the temperature drops to 59°F (15°C), which is higher than the European adder encounters during most of its short Arctic summer. The optimum temperature for rattlesnakes is 77°–89°F (25°–32°C).

Snakes also estivate. In the extreme summer temperatures of temperate zone arid regions, such as Death Valley, snakes take refuge from the heat. The water moccasin or cottonmouth, an aquatic, highly venomous snake of the eastern United States, is said to estivate if its habitat dries up in midsummer, when it finds a cool and protected cavity and awaits the return of the rains.

Some of the Species

Colubrids

The large and widespread family *Colubridae* contains almost 80 percent of all the living snakes, and its members form an even larger percentage of the snakes of some regions, especially North America. Colubrids are distributed throughout the world, except the polar zones, and are just as wide-ranging in habitat and behavior. The family contains both nonvenomous forms (aglyphs), which have simple teeth, and mildly venomous species (opisthoglyphs). The latter do not have hollow fangs, but are "rear-fanged" and chew on their prey to transmit venom through grooved maxillary teeth. Colubrid snakes bear no traces of hindlimbs like the boas and pythons, and the left lung is either very small or absent. The family contains both egg-layers and snakes that bear live young; it includes many northern temperate species, and therefore many snakes that must hibernate to survive the cold season. These include the garter snakes—the northern-most reptiles in the Nearctic Zoogeographic Region—and in the Palaearctic Region such snakes as the large whipsnake (*Coluber jugularis*), at 10 feet (3 m) long the largest European species, which ranges from southeastern Europe to central Asia; and the 6-foot-(1.8 m) long four-lined snake (*Elaphe quatuorlineata*) with a similar distribution, both of which hibernate from November to March. Elevation is also a factor in the length of hibernation, and snakes that live high on mountain slopes, even in the milder parts of the temperate zone, hibernate for at least half of the year. The southern smooth snake (*Coronella girondica*), although mainly a reptile of the Mediterranean regions including northwest Africa, occurs up to 10,500 feet (3,200 m) and must hibernate from early October to April, as does the Eurasian rat snake (*Elaphe dione*) of central and eastern Asia from the Ukraine to the Pacific Ocean, which has been found at up to 11,500 feet (3,500 m).

Even snakes living in the milder regions around the Mediterranean Sea also hibernate. The tesselated snake (*Natrix tesselata*), an aquatic fish-eater of southeastern Europe, Asia Minor, and the Middle East, hibernates in holes near the water's edge from October to March, and the leopard snake (*Elaphe situla*) of southern Italy and the Adriatic coast of Bosnia and Croatia becomes torpid in late October and does not emerge until March.

Red-sided Garter Snake (*Thamnophis sirtalis parietalis*)

This is a very cold-hardy snake that occurs in the center of the continent from northern Canada (in the Northwest Territories just south of the Arctic Circle) south to Oklahoma. In general color it is olive-brown or black with alternating red- or orange-and-black bars along the sides of the body, giving the appearance of a red stripe, between lateral yellow stripes. Old females average 3 feet (90 cm) in length, and males 30 inches (76 cm). It is a live-bearer, which produces an average of twenty young per litter after a gestation period of 12–14 weeks, but sperm retention can result in births many months after mating, even occasionally in the early summer of the following year. In the northern parts of its range, and especially in Manitoba's Interlake region, this snake is known worldwide for its habit of hibernating communally. Several thousand may congregate in traditional hibernacula in limestone sink-holes, which has made them vulnerable to modern-day

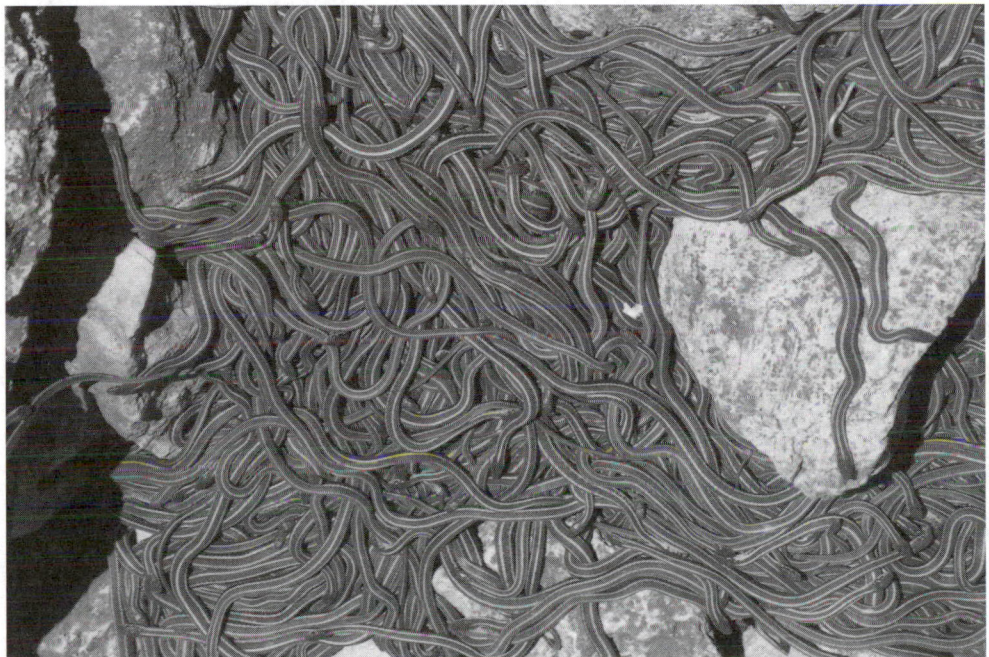

Red-sided Garter Snakes *Famous internationally for their habit of congregating in large numbers at traditional hibernation sites, red-sided garter snakes mass outside their dens in early spring, warming up and mating. They return to their underground dens for the night until it is warm enough to migrate to the sloughs and marshes where they feed on frogs.*
Photo: Clive Roots

hazards such as collecting for the pet trade and climate change. They migrate back to their hibernacula in early September, but stay out during the day sunning themselves until cool and wet weather finally forces them underground for the winter. At the end of April they come out to bask at the edge of the dens in the weak sunlight, going back down again at night. Mating occurs during this period, with many males chasing a female, even into the branches of the shrubs around the dens. At the end of May they migrate out into the lakes and sloughs to feed, mainly on frogs, leeches, salamanders, and earthworms.

Smooth Snake (*Coronella austriaca*)

The smooth snake is a slender, nonvenomous species of variable color, usually gray, brown, or reddish-brown with a double row of brown or black spots along the back, these sometimes joining to form two stripes, and although it bears a strong resemblance to the adder, it is distinguished by its rounder head and long, slender build. It rarely exceeds 24 inches (61 cm) in length, and is a timid and gentle snake that prefers dry and sunny areas, including scree slopes, heathland, embankments, and dry open woodland. The smooth snake occurs across Europe as far north as central Norway, and eastward to southwestern Asia. It is mainly a lizard- and snake-eater, plus mice and voles, and is a constrictor that squeezes its prey in its coils and is able to overpower and then swallow snakes almost its own length. In central Europe the smooth snake hibernates communally in October and reappears in early April. It is a viviparous species, its young being born alive but covered by a thin membrane that tears immediately when they are expelled; this form of reptile reproduction most closely resembles the nourishment of the young within the mother's body by the placental mammals.

Grass Snake (*Natrix natrix*)

This is a common nonvenomous snake of the British Isles and Europe (although absent from Ireland, Malta, Crete, and the Balearic Islands), and northwest Africa, Asia Minor, and eastward across Asia to Lake Baikal. In Scandinavia it extends northward to northern Sweden, and has been found at up to 7,500 feet (2,300 m) in the Italian Alps. It is a thick-bodied snake when adult, and females are larger than males, averaging about 5 feet (1.5 m), but reaching a length of 6 feet 6 inches (2 m) in Sardinia and Sicily. Although variable in color, grass snakes are usually olive-gray or greenish-brown, and most individuals have a yellow collar bordered with black. Diurnal in its habits, it prefers dry chalk hills, woodland, farmland, and grasslands, often some distance from water. The grass snake lays up to forty-eight eggs, often in piles of farmyard manure (which were once a common feature of the British countryside), where they are incubated by the heat of decomposition. It is quite catholic in its prey selection, eating fish and small rodents, plus frogs, toads, newts, and salamanders in both larval and adult form. Wild specimens are usually retaliatory and aggressive when handled, and have the unpleasant habit of discharging their cloacal glands. Grass snakes hibernate from October to March or early April in the British Isles and northern Europe in rabbit holes, deep rock crevices, and manure piles.

Northern Water Snake (*Natrix sipedon*)

The northern water snake occurs throughout eastern North America, excluding the southeastern corner, from Louisiana north into southern Canada. A stout-bodied snake, it reaches a length of 4 feet (1.2 m) and is the only large water snake in the northern parts of its range; it is a very common species, found in just about every body of fresh water. More aquatic than the grass snake, it prefers still or slow-moving water, where it can remain submerged for up to an hour. The northern water snake is generally pale-gray to brown with darker cross bands and blotches, and has a yellowish-orange belly; but its coloration often darkens with age, obscuring the markings and appearing solid dark brown. Like all water snakes it strikes and bites hard when handled, and is also disliked because of its natural predation on fish. It is active in daylight and at night and loves to sunbathe on a bank or rock at the water's edge. It hibernates colonially under piles of debris or in holes above the water line, from late October to April in southern Canada. Like the grass snake it discharges a foul-smelling musk to deter predators, which include hawks, snapping turtles, foxes, raccoons, and opossums.

Plains Hognose Snake (*Heterodon n. nasicus*)

A snake of the prairies from southern Canada to Mexico, the plains hognose is a slow and clumsy terrestrial reptile, with an upturned hard scale at the end of its snout, an adaptation for burrowing and searching for the toads that are its main prey, and large teeth for holding them. Consequently, it prefers the dry, sandy, and gravelly terrain of well-drained grassland, dry woods, and fields, which are toad habitat. It is mildly venomous but is not considered dangerous as it can hardly be induced to bite, and has evolved highly unusual behavior for when it is threatened. At first it hides its head under its coils, but if that tactic fails then it inflates itself with air, spreads its jaws and neck and hisses in imitation of a cobra. If further deterrent is needed it rolls onto its back and very effectively "plays dead," with its mouth open and tongue lolling out, but if it is righted it immediately turns back over again, giving the game away. The hognose snake is a heavy-bodied animal that reaches a length of 3 feet (90 cm). It is brown with darker-brown blotches along the back, two alternating rows of small dark spots on the sides, and a large blotch on the side of the neck; and the ventral surface is shiny black with small yellowish-white squarish markings. It hibernates from September to March in southern Canada, usually in the holes of ground squirrels. In addition to toads it eats frogs, small snakes, mice, voles, and young birds.

Black Rat Snake (*Elaphe o. obsoleta*)

The black rat snake is a glossy black serpent with a white chin and throat and a grayish-yellow belly with black "checkerboard" markings, and is typically about 6 feet (1.8 m) long but may reach 8 feet (2.5 m). It has a wide range through the eastern and central parts of North America from southern Ontario to Wisconsin and south to Oklahoma and Louisiana, but does not reach the Gulf of Mexico nor the southeastern coastal areas and Florida. Forests, marshes, farmland, and grassland are all included in its wide choice of habitat, where it is diurnal in the spring and fall but nocturnal during the summer months. A powerful nonvenomous constrictor, often killed simply because its size implies that it is dangerous, as a rat-killer it is considered a highly beneficial species, especially as it is a good climber that searches the rafters of farm buildings for rodents and bats. It also eats mice, voles, young ground squirrels, chipmunks, frogs, toads, and other snakes, and goes high into trees to raid birds' nests. The black rat snake is itself a victim of raccoons, foxes, bobcats, owls, and hawks. In October in Minnesota and Wisconsin, it hibernates communally in dens, in natural rock fissures and faults, and in decayed root systems, and does not appear again until the end of March.

Aesculapian Snake (*Elaphe longissima*)

The Aesculapian snake is a European relative of the American rat snake and is a long and slender reptile with smooth and flat keeled scales; it reaches an average length of 5 feet 6 inches (1.7 m). Its range is central and southeastern Europe (from Denmark to Italy) and then eastward into Asia Minor and Transcaucasia. In color and pattern it shows the great variability so common in rat snakes, but most individuals are gray-brown or olive-brown with small white spots at the edges of the scales and a pale yellowish-white belly. Adult snakes often have just faint stripes and no spots and are then mistaken for four-lined snakes; their young, which are spotted and have yellow on the sides of the head, are often confused with grass snakes, but their scales are not keeled. Aesculapian snakes are good climbers, and prefer dry habitat, especially sheltered and sunny scrubland and open woodland, where they seek mice and voles, birds and lizards. They are very sensitive to cold; they begin their hibernation in early October and may not emerge until early May. They also avoid excessive heat and remain hidden during the day in midsummer.

Corn Snake (*Elaphe guttata guttata*)

The corn snake is the most colorful of the rat snakes, with considerable variation in color, but generally with reddish blotches bordered with black on a pale yellowish background; it is also called the red rat snake. It is one of the smaller species, averaging 39 inches (1 m) in length, although individuals up to 70 inches (1.8 m) have been recorded. It has a wide distribution across the southern United States and into northern Mexico, where it prefers drier habitat such as dry river bottoms, rocky hillsides, and pine barrens. Like the other rat snakes it is mainly terrestrial but is a good climber. The most popular snakes kept as pets, corn snakes have responded to improved breeding techniques and selective breeding by producing many attractive mutations. There has been extensive production of color morphs since the first albino was bred in captivity in 1962, and each year U.S. breeders produce thousands of amelanistic corn snakes in which the lack of normal black pigmentation results in numerous mutations ranging from blood-red to white.

Bull Snake (*Pituophis melanoleucus sayi*)

Another snake of the central grasslands of North America, the bull snake prefers dry and sandy areas on the prairies, open woodland, coulees, and river bluffs, from southern Alberta and Saskatchewan south to Texas. It is a large and powerful, heavy-bodied snake that averages 6 feet (1.8 m) in length, although the record is 8 feet 4 inches (2.5 m). The bull snake has keeled scales that have longitudinal ridges on them, and an overhanging (supraocular) scale above each eye, which gives it a "scowling" appearance. It has a small head and pointed snout, and normally has a yellowish or tan base color, with black or reddish-brown blotches on the back and sides, and a yellow belly with black spots. The bull snake constricts its prey, and like the rat snakes, hisses, raises its head, vibrates its tail, and strikes when startled or cornered, but it is not venomous. Sun-warmed rocks and metalled roads are favorite warming-up places, and it is mainly terrestrial and diurnal. Rodents are its favored prey and the snake follows them down their holes and presses them against the side of the burrow with its body while seizing one at a time. It also eats lizards and birds. On the northern plains, it hibernates in rodent burrows from early October to March.

King Snakes (*Lampropeltis*)

King snakes are powerful cannibalistic constrictors that prey upon other snakes including venomous species to whose venom they are apparently immune; but they also eat lizards, birds and their eggs, plus those of other reptiles. They reach an average length of about 48 inches (1.2 m), with the record specimen—an eastern king snake—measuring 78 inches (2 m) long. They are mainly heavily patterned snakes with dark-brown blotches on a brown background, but with such a wide distribution across North and Central America, there are several species and races. The common kingsnake (*Lampropeltis getulus*) has a wide range across the southern half of the United States. The type race known as the eastern kingsnake (*L. g. getulus*) is a mainly diurnal subspecies that prefers wetlands where it preys largely on water snakes and turtle eggs. It is a shiny-black snake with bold white "chain-link" markings. The desert kingsnake (*L. g. splendida*) lives in the arid Southwest where it has assumed more nocturnal habits to escape the daytime heat. It is completely spotted with whitish-yellow dots along its sides and has a vertebral line of larger dark-brown spots. The gray-banded king snake (*L. alterna*) (see color insert) from Texas has variable coloration from dark gray to light gray with wide red or orange bands or "saddles" bordered on each side by fine white bands and wider black bands. It is not as food-specific and eats small rodents, frogs, and lizards.

Vipers

The members of the family *Viperidae* are the only snakes with hollow, hinged fangs at the front of the upper jawbone, and when not in use they are folded back against the roof of the mouth to allow it to close. Consequently, their fangs can grow quite large, and venom from their glands—which are modified salivary glands—can be injected quickly by the large muscles at the back of the head. These swing the fangs down and forward when the snake opens its mouth, and the venom is discharged the moment the fangs enter the victim. They are very dangerous snakes, their venom acting quickly on the prey's vascular system, destroying blood cells and causing hemorrhaging.

The vipers are ambush predators that strike and then follow the scent of their stricken prey. They are generally sedentary and well camouflaged, and lie in wait until their prey comes within reach; they can tell from the intensity of the ground vibrations whether the approaching animal is large and therefore a potential threat, when they lie still or slip quietly away. After striking their prey and injecting venom, they trail it by means of their Jacobsen's organs until it succumbs, which is after just a few seconds for a small rodent, as the hemotoxic venom acts very quickly. Members of the family occur throughout the world except the polar regions, Madagascar, Australia, and New Zealand, and are divided into two subfamilies which separate the vipers or adders (*Viperinae*) from the pit vipers (*Crotalinae*).

The *Viperinae* subfamily includes many Old World species such as the stout-bodied venomous snakes of Africa—the rhinoceros viper (*Bitis nasicornis*), puff adder (*Bitis arietans*), and Gaboon viper (*Bitis gabonica*). The puff adder is very sensitive to cold and hibernates during the South African winter, and the numerous vipers that occur in temperate Eurasia are also forced to hibernate. These include the adder (*Vipera berus*), Britain's only venomous snake, and the blunt-nosed viper (*Vipera lebertina*) of Europe, central Asia, and North Africa—the largest species with a length of 6 feet 6 inches (2 m)—which hibernates from October to March even in the milder latitudes of its range, but emerges to bask when the temperature reaches 48.2°F (9°C). Others that must hibernate are the nose-horned viper (*Vipera ammodytes*) of southeastern Europe and Orsini's viper (*Viper ursinii*), which ranges from eastern Europe to central Asia.

The *Crotalinae* subfamily includes the more familiar rattlesnakes (*Crotalus*), the copperhead (*Agkistrodon contortrix*), and the water moccasin or cottonmouth (*Agkistrodon piscivorus*), which are the most specialized of all snakes, adapted for hunting in total darkness. They are widespread throughout the warmer regions of the world, except Africa and Australia, and are ecologically adapted for life in a wide range of habitat. There are terrestrial rain forest species such as the 10-foot-long Bushmaster, several prehensile-tailed arboreal snakes of which the 4-foot-long Nicobar Island pit viper is the largest, and desert dwellers like the sidewinder (*Crotalus ceraste*). Most pit vipers give birth to live young, with the bushmaster being the only New World species to lay eggs, and in the Old World the Malayan pit viper (*Agkistrodon rhodostoma*) is also an egg-layer.

The sinister appearance of the pit vipers heightens people's general apprehension of snakes and their hostility toward them. In common with other snakes they have the typical "cold," expressionless eyes and of course lack eyelids, but their sinister looks are enhanced by their large, triangular-shaped heads, vertical pupils, and the enlarged scale that hangs over the eye. They have almost unequalled killing power, and many people around the world succumb annually to their long and deeply penetrating fangs and the large amount of venom they can inject quickly, which breaks down the walls of the blood vessels and the leucocytes, although there are specific variations. The South American rattlesnake, pygmy rattlesnake, and the Western Mexican rattlesnake are especially dangerous as their venom also contains a neurotoxic element that acts on the nervous system, so a special antivenin is needed to treat their bite. The digestive enzymes present in the venom of the eastern diamondback rattlesnake have a necrotizing effect on flesh, and digits or limbs have been lost from the effects of its bite. Fer de lance venom has a very spectacular effect on its victims, causing hemorrhaging, with blood oozing out of all the body's orifices.

Like the *Boidae*, which have heat sensors on the lips, the *Crotalinae* also have heat-sensing adaptations but they are far more sophisticated. They have a sensory loreal or facial pit, about 0.5 cm deep, below and in front of the eye on each side of the head, almost forming a triangle with the eye and the nostril. The pit is a cavity in the maxillary bone, covered with a membrane and connected to the outside by a tiny hole. The membrane is well supplied with 7,000 nerve endings and receptor organs sensitive to the infrared radiation (invisible rays beyond the red end of the spectrum) whose wavelengths are too long to stimulate the human retina, and too short to produce the stimulation of heat in human receptors. They are sensitive to temperature differences as low as 0.02°C, and can accurately locate prey animals by the heat emanating from them. On a night when the ambient temperature is less than that of a rat kangaroo, a rattlesnake can detect the rodent 18 inches (46 cm) away by its body temperature. To prove that the pit is a true sense organ, which other animals do not possess, researchers blindfolded rattlesnakes, removed their tongues and plugged their nostrils. The snakes struck accurately at covered light bulbs, confirming that their pits guided them to the source of heat.

Rattlesnakes are the most numerous species of pit vipers in the New World, occurring from southern Canada to Argentina. Composed of horny, loosely interlocking segments at the end of the snake's tail the rattle functions as a warning only. Newborn rattlers have a single button on their tails, and add a segment every time they shed their skins, which is determined by their growth rate. This occurs faster in young snakes, which may add up to four rattle segments annually, slowing down to one per year as they mature. Rattlesnakes have colonized a wide range of terrestrial habitats, from grassland and desert to dense woodland. The sidewinder is a desert species, its name derived from its peculiar sideways locomotion in soft and loose sand, where other snakes do not venture. Progression in this way is very swift and also reduces the snake's body contact with the hot sand, although it normally hides during the day in a rodent burrow and comes out at night to seek kangaroo rats.

The prairie rattler also hides during the day in a burrow, usually a prairie dog's, even an occupied one, and young prairie dogs and burrowing owls that share the same tunnel systems are a handy source of food for the rattlesnakes. In the northern parts of its range the prairie rattlesnake encounters short summers and low winter temperatures, which result in slowing the growth rate of the young, so that births occur every other year instead of annually. In North America all rattlesnakes living in localities where the temperature drops below 60°F (16.0°C) hibernate in winter, the period varying according to the length of the cold season.

Asp Viper (*Vipera aspis*)

The asp viper is a more elongated snake than the adder, and its snout is turned up at the tip. Its basic ground color is very variable, and individuals may be gray, brown, reddish-brown, brownish-orange, or coppery-red, with narrow black cross-bars on the back alternating with similar markings on the sides, and all connected by a narrow dark line along the spine. The underparts are also variable, but are usually black or dark blue. Its maximum length is 2 feet 6 inches (76 cm). The asp viper's bite may be fatal to man, especially small children, and if the snake is a big one, able to inject a large amount of venom. Its range is central and southern Europe from the Mediterranean region north to Sweden. In the Pyrenees, it has been seen at an elevation of 7,000 feet (2,135 m) and in the Alps even higher, up to 9,700 feet (2,960 m). It prefers hot and dry habitat, rocky scrubland, and dry forests, where it hunts mice, voles, young birds, and lizards. In France it hibernates in late October and emerges at the end of March, and occasionally in early February depending upon the temperature. It also appears on warm days in midwinter to bask in the sun. Hibernation sites are holes in the soil or deep rock crevices. The snake that supposedly killed Cleopatra was an asp or Egyptian cobra, not an asp viper.

European Adder or Viper (*Vipera berus*)

The most widespread venomous snake in Europe, the adder occurs in all countries except Ireland and is the only venomous reptile in Britain, although its bite has rarely been fatal to humans. It is also found westward across northern Asia to the Pacific Ocean, and has the most northerly range of all snakes, living just inside the Arctic Circle in Siberia, just south of the range limit of the viviparous lizard, and therefore the world's second-northern-most reptile. In Switzerland it lives at up to 9,000 feet (2,745 m) in the Alps. Throughout its range the adder hibernates during the winter months, but despite being mainly nocturnal it must bask frequently to warm up, especially during the weak sunshine of spring and autumn. In the far north the short summers restrict its active season to a maximum of four months, and it hibernates in deep crevices and sink-holes to stay below the frost, which penetrates at least 6 feet (1.8 m). The short season is hard on females, for they must mate, complete the internal incubation period of their eggs, and give birth to live young, which must then have time to store fat before the arrival of cold weather; many do not survive their first winter. The adder prefers dry moors, sandy heaths and dry woodland, and hillsides exposed to the sun; its prey are mice, voles and shrews, frogs, toads, and lizards. The male's body color is creamy-yellow to gray and the female's reddish-brown, both with a dark zig-zag pattern along the spine, and the maximum length for the species is about 24 inches (61 cm).

Western Diamondback Rattlesnake (*Crotalus atrox*)

The "diamondback" is the largest "rattler" in western North America, occurring in the southwestern United States and adjoining northern Mexico. It is a thick-bodied snake with keeled scales, and has the triangular head and narrow neck typical of the vipers; it reaches a length of about 7 feet (2.10 m). Varying in color from gray or yellow to dark brown, it has light-brown to black diamond-shaped blotches that fade toward the tail. Its tail is ringed with black and white and tipped with a rattle. It is a stubborn snake that usually stands its ground when threatened, but should be left strictly alone as its bite is potentially deadly unless medical treatment is immediately available. Rattlesnake venom causes extensive tissue damage and internal bleeding, and retains its qualities for long periods, being still dangerous when fifty years old. Its toxicity is compounded by the large amount of venom injected, for the glands of a mature diamondback can hold 600 mg of venom. A nocturnal species, the diamondback is active by day in the coolness of spring, and basks in the late afternoon. It is largely a rodent-eater, preying on

Western Diamondback Rattlesnake *A "rattler" of the arid southwest, its large size (up to 7'), the capacity of its venom glands, and the encroachment of humans into its habitat make it the most dangerous snake in North America. It hibernates communally in caves or rock crevices from October to March at the higher elevations of its range. Its heat-sensing pits can clearly be seen between and below the eyes and nostrils.*
Photo: Courtesy Patrick Jean, Nantes Museum

kangaroo rats, rabbits, and ground squirrels, whose burrows it enters. It is viviparous and gives birth to about twenty young in late summer, which have a single button on their tails and add a segment each time they shed their skins. The diamondback hibernates communally, sometimes in the hundreds, in traditional sites such as caves and deep rock crevices.

Eastern Massassauga (*Sistrurus c. caternatus*)

This was the common rattlesnake of southeastern Canada and neighboring New York State, from where it ranges in a southwesterly direction to Iowa and Missouri, but it is now rare in many places due to persecution and the loss of habitat to agriculture. It is also known as the swamp rattler because of its preference for moist habitat, including grassland, sedge meadows, bogs, and swamps, but it avoids open water and forests. The massassauga is a small snake, reaching a length of only 2 feet 6 inches (76 cm), and is either gray or grayish-brown in color with rows of darker spots—a row of large spots along the spine and three rows of smaller spots on each side—and it has a black belly. It has the typical triangular rattlesnake head with black stripes. Unlike rattlers generally, it is not an aggressive reptile, and must be really annoyed before it rattles. Its diet includes frogs, mice, rats, birds, and smaller snakes; and it hides, and hibernates, underground, in the Midwest from late September to the end of April, in mammal or crayfish holes or natural crevices.

Timber Rattlesnake (*Crotalus horridus*)

A heavy-bodied snake when adult and with a large triangular head, the timber rattlesnake lives in the mixed coniferous and deciduous woodland in the eastern United States from southern New Hampshire to northern Georgia and westward to Wisconsin and Texas. It prefers rugged terrain with exposed rock ledges on which it can bask, but is disappearing over much of its range due to the loss of habitat to development, den destruction, and hunting. It generally reaches a length of about 48 inches (1.2 m) but occasionally individuals have grown to just over 72 inches (1.8 m). It is a very slow-growing snake, especially in the north of its range where summers are short, and it may not reach maturity until eleven years old and may then live until the age of thirty years. It is a very slow breeder also, laying eggs every three or four years, which develop completely within the female enclosed in a thin membranous "eggshell" which ruptures almost immediately when they are laid. The hatchlings have usable fangs and venom and one tiny rattle segment called a "button." Thereafter, they grow a new rattle each time they shed their skins, which in this species happens only once every year to eighteen months. There are two color phases of the timber rattlesnake—a yellow one in which the base color is yellowish-brown with dark-brown or black crossbands, and a darker one with the same crossbands on a dark, almost black background.

Mature timber rattlers have few natural enemies, but the young are vulnerable to a number of predators, including hawks and owls, coyotes, foxes, and raccoons.

Their main prey animals are rodents, small rabbits, opossums, and ground squirrels. For their winter hibernation, timber rattlesnakes den communally in groups of up to 100 in a rock crevice or cave below the frost line, often in company with other snakes. In Wisconsin and New York they are torpid from mid-October to mid-May.

■ TURTLES

In North America the aquatic turtles and terrapins and the terrestrial tortoises and box turtles are referred to collectively as turtles, whereas in Europe the name "turtle" is normally used only for aquatic species. They are all members of the order *Testudinata* which is synonymous with *Chelonia*, and to avoid confusion they are often all called chelonians. Turtles are the most primitive of the reptiles, and are believed to have arisen at least 250 million years ago. They now occur throughout the world in the tropics and subtropics, and in temperate latitudes where they avoid extremes by hibernating in winter and estivating in summer.

The most obvious characteristic of all turtles is their bony two-part shell with limbs, head, and tail protruding from between them. The upper shell is called the carapace and the lower shell the plastron, and they are formed by bony plates that are the shell's structural components, which are covered by horny scutes and in some species by skin. The shell is not an exoskeleton or external skeleton, however, but a modified rib cage and part of the spinal column, with the bony plates of the carapace being fused to the ribs and the trunk vertebrae. The two sections of shell are attached at the sides by bony structures called bridges. The turtle's limb bones are modified and the pectoral and pelvic girdles are attached to the inside of the shell. Turtles vary in their external covering, which is modified epidermis, from the unsculptured shell of the diamondback terrapin (*Malaclemys terrapin*) to the large horny scutes of the land tortoises, and the leathery and often spiky carapaces of the soft-shelled turtles. The shells also vary considerably in shape; in the land tortoises from the dorso-ventrally flattened carapace of the pancake tortoise to the high-domed, and sometimes saddle-backed, shapes of the giant tortoises; and in aquatic species from the flattened shells of the soft-shelled turtles to the heavily ridged carapace of the alligator snapper. The carapace is always a single shell, but in some species the plastron is segmented and hinged, allowing the animal to withdraw its head, legs, and tail and close up tightly. Males generally have concave plastrons that fit over the female's rounded carapace and aid mating. Turtles protect themselves from danger by withdrawing their heads, and in some species their legs and tails also go into their shells.

Turtles lack teeth but most have powerful cutting and crushing jaws, which are covered with a horny sheath similar to a bird's beak, and with either a sharp or serrated edge. They vary in size from turtles just 3 inches (8 cm) long to the giant land tortoises of the Galapagos and Aldabra Islands which can weigh 350 pounds (158 kg), and the huge freshwater turtles such as the alligator snapper (*Macroclemys temminckii*) weighing up to 220 pounds (100 kg). Unlike the snakes, which shed their whole skin at one time, shedding in the chelonians is a continual process with pieces of the scutes continually flaking off. Turtles are long-lived, especially the giant land tortoises, which frequently become centenarians. A long lifespan is not necessarily dependent upon large size, however, for even the tiny musk turtles have lived for almost half a century, and the desert tortoise may live a century. There are fewer examples of confirmed lifespans where the aquatic species are concerned, but the huge size reached by the alligator snapper is a sure sign of long life. Allied to this longevity is their relatively slow growth rate, and sexual maturity is rarely attained until the age of ten years in many species.

Coupled with their slow reproductive rate the genetic investment in a mature tortoise or turtle is obviously tremendous and has resulted in their ability to survive for millions of years. Unfortunately, it has not prepared them for the rapid losses to their numbers through the onslaught of the modern age, and the demand by pet shops and restaurants. The great demand for pets led to the large-scale production of farmed turtles, particularly in the southeastern United States, but as specimens under 4 inches (10 cm) in length cannot be sold in the United States, due to the risk of salmonellosis, they are now exported to Asia as pets and potential food animals.

Hibernation

Turtles do not range as far into the higher latitudes of the Northern Hemisphere as the snakes and lizards. In Europe the terrestrial tortoises are mainly animals of the countries bordering the Mediterranean. Some extend eastward into Iran and Horsfield's tortoise (*Testudo horsefieldi*) ranges north of the Caspian Sea; the distribution of the aquatic European pond terrapin (*Emys orbicularis*) extends only to southern Germany and Poland. In Asia, Reeves terrapin (*Chinemys reevesi*) and the Chinese softshell (*Trionyx sinensis*) occur farther north—in Korea, northern China, and Japan. In North America the painted turtle (*Chrysemys picta*) is the northern-most aquatic species, occurring in Lakes Winnipeg and Manitoba in the eastern prairies, while the virtually terrestrial eastern box turtle (*Terrapene c. carolina*) occurs in the northeastern United States almost to the Canadian border.

Most turtles hibernate when the temperature is between 38°F (3.5°C) and 44.6°F (7.0°C), so wherever they live, whether in Alabama, Italy, or Honshu, when it cools to those levels they will become torpid. Land turtles may hibernate under a shallow covering of soil like the box turtles, or may just push beneath a thick shrub and stay there semi-exposed to the cold like the Texas tortoise (*Gopherus berlandieri*); or they burrow deep like the desert tortoise (*Gopherus agassizi*). There is a subtle difference between torpor and hibernation which is very important to the turtle in winter. They may become torpid at about 50°F (10°C), but at that temperature their metabolism is not lowered to the level of normal hibernation, and in this midway "twilight" zone their fat stores will not last so they would starve to death before the end of winter. Captive animals should have the lowered temperature needed for full hibernation, or they must be kept sufficiently warm to be active and able to feed. In preparation for dormancy they stop eating as the time for hibernation draws near, to allow their digestive systems time to clear, since their metabolism virtually shuts down and cannot cope with undigested food in the gut. Burrowing species select soil that is slightly moist for their tunnels, as this reduces the risk of evaporative water loss through their skins and rapid dehydration.

Aquatic turtles hibernate on the bottom of their pond, either totally exposed or under debris, or buried in the mud with just their heads exposed. The hibernating turtle's metabolism is so slow it can get sufficient oxygen from the meager amount in the mud. The oxygen is absorbed through the mucous membrane of their throat, and through thin-walled sacs in the cloaca called bursa. However, their oxygen requirement is considerably reduced to provide the minimal needs of their lowered metabolism, and metabolic adaptations allow them to survive without oxygen for several months. Painted turtles and pond sliders have the greatest tolerance for total oxygen deprivation, and the oxygen in their blood can fall to almost zero for four months without causing harm. To provide the energy needed for their lowered metabolism, turtles store food for winter as lipids, which occur as solid fatty masses in the viscera, under the skin and in the muscle tissue; and as glycogen, which is stored in the liver and tissues. They are able to survive without oxygen because the glycogen is broken down by the process of glycolysis which does not need oxygen. The lactic acid produced by this process, which would normally be toxic, is buffered by calcium and magnesium drawn from the turtle's bones and shell. Anaerobic survival is also aided by the hibernating turtle's ability to shut down its metabolism to about 10 percent of its normal rate, and so extend the life of its reserves.

Like the northern, partially freeze-tolerant frogs, at least two turtles can tolerate the freezing of their extracellular fluids. Both live in northern latitudes and neither have evolved behavior to avoid freezing temperatures, but tolerate them instead. The eastern box turtle (*Terrapene c. carolina*) hibernates under a thin layer of leaves and soil, just like the wood frog, but this does not give adequate protection against frost, and it can withstand the freezing of almost 60 percent of its body water to temperatures down to 25.4°F (−3.6°C) for several days without injury. Short summers at the northern end of the range of the painted turtle (*Chrysemys picta*) leave insufficient time for the hatchlings to feed and lay down stores of fat for hibernation, so they stay where they hatch, under perhaps 1 foot (30 cm) of sandy soil, until the following spring. In zones where the frost can penetrate as much as 6 feet (1.8 m), which applies to most of their range in central Canada, the hatchling turtles obviously freeze, yet they survive to emerge the following spring because they have natural freeze protection. They have no heartbeat, muscle movement, or blood flow, and very minimal brain activity, but they are not actually frozen solid, which would be fatal. It is their extracellular fluids such as blood and urine that have frozen without harm, due to their ability to produce cryoprotectants or "antifreeze" carbohydrate compounds that protect the cell contents.

Estivation

The distribution of the turtles also includes many regions where they must escape from extremely hot and dry conditions, in temperate regions as well as in the tropics, and like many other reptiles and amphibians their behavioral response to such conditions and the drying up of their ponds and rivers is to estivate. This behavior is triggered by increasing temperatures and drought and is ended by rain and lowered temperatures. Aquatic species lie dormant in the dried-out mud of their pond until the rains revive them. When their pond water evaporates completely the short-necked turtles (*Emydura*) of Australia leave their pool and estivate on land, burying themselves in damp soil in a spot sheltered from the direct sun. In North America the Sonoran mud turtle (*Kinosternon sonorae*) of southern Arizona, New Mexico, and adjacent Mexico estivates for about a month when its pond dries out in late summer. The western chicken turtle (*Deirochelys reticulata*) moves away from its dried-out pond and estivates buried in leaves, and the red-bellied turtle (*Pseudemys rubriventris*) digs into the mud at the bottom of the pond, leaving just its nostrils above the surface.

The terrestrial species hide from the heat in their burrows. In Africa both the spurred tortoise (*Geochelone sulcata*) and the leopard tortoise (*Geochelone pardalis*), which live in the sub-Saharan savannah and thorn scrub, estivate for several weeks underground. In North America the desert tortoise (*Gopherus agassizii*) of the Sonoran Desert stays in its burrow to avoid the midsummer heat, but the Texas tortoise (*Gopherus berlandieri*), also a burrower, withstands extreme heat and dryness better than the previous species, and is therefore more inclined to make a scrape under a bush or clump of grass to estivate. As they must also hibernate to escape the cold of winter, these species are inactive for almost eight months each year. This also applies to other temperate desert-dwellers, such as Horsfield's tortoise (*Testudo horsfieldi*), which hibernates and estivates for almost eight months of each year in central Asia; and the desert box turtle (*Terrapene ornata luteola*), which spends three months estivating in midsummer soon after its five months' winter hibernation.

Some of the Species

Freshwater Turtles

These are the turtles and terrapins that are adapted for life in inland freshwaters, and have either a complete or semicomplete aquatic existence. Most have long toes connected by webbing—for increased thrust in the water—and flattened, smooth shells. They can walk on the bottom of their pond, and the soft-shelled species swim almost as fast as fish. They are widely distributed across the world, and are particularly well represented in the New World. Many are temperate-zone animals and must therefore become torpid seasonally to avoid low water temperatures, and most species hibernate in the water, which is a much safer environment since they are never exposed to the extremes they would experience above ground. Beneath the ice the water remains just above freezing, and if there is a risk of the ice reaching them they move to deeper water. However, these turtles are totally dependent upon land for incubating their eggs and they also haul out of the water daily to sunbathe and raise their body temperature quickly. They are mainly carnivorous, although some may also eat plant matter. Like the land turtles they are air-breathers and must therefore surface during normal (nonhibernating) activity. As their ribs are now part of their shell they cannot use them to draw air into the lungs, and they alternately expand and contract muscles in the abdomen and above the limbs to change the space within the shell to inhale and exhale.

Painted Turtle (*Chrysemys picta*)

This turtle has a very wide range in North America, from coast to coast across the continent, down to the Mississippi Delta and extending northward into southern Canada and into Manitoba's Lake Winnipeg and Lake Manitoba, the highest (and coldest) latitude for turtles in North America.[9] It is rarely more than 8 inches (20 cm) long, and has a low, smooth, and unkeeled carapace, which is usually dark green with red markings on the marginal scutes. The plastron is yellowish-orange with a long gray blotch down the middle, and the head and neck are distinctively striped with red, yellow, or orange lines. The painted turtle prefers the still or sluggish water of ponds, lakes, marshes, ditches, and prairie sloughs with muddy bottoms and profuse plant growth, and it basks on logs and mudbanks, often communally. Omnivorous, its diet includes aquatic plants and insects, crayfish, frogs, tadpoles, and snails, and it also scavenges on carrion. Its clutch of eight eggs is laid in mid-summer in a hole scooped out in the south-facing sandy bank of a pond or lake and is left to incubate by radiation. The eggs take about ten weeks to hatch and the hatchlings hibernate where they hatch, and are partially frozen, surviving only because they can produce cryoprotectants or "anti-freeze" compounds which protect the cell contents from freezing. From their second winter onward, however, they hibernate underwater in typical aquatic turtle manner, usually resting on the mud at the bottom of the pond or lake. The painted turtle prefers shallow water for hibernation, moving farther out as the ice thickens, but does not go to the center of large lakes where the great ice sheet over a large surface takes longer to thaw in spring and delays its emergence. Hibernation in the north lasts from October to April, this period shortening at lower latitudes, until eventually not at all in the Mississippi Delta.

Red-eared Turtle or Slider (*Chrysemys scripta*)

A greenish-yellow turtle about 10 inches (25 cm) long, the red-eared slider is a very recognizable species as it is the only North American turtle with a broad red stripe behind the eye, in the males at least, as the female's stripe may be indistinct. However, the stripe is sometimes completely lacking in both sexes and in some individuals it is yellow. The slider is also the most popular pet turtle, although not to the same extent since the ban on the sale of small specimens, due mainly to the turtle farms in the southeastern United States being considered inhumane and unsanitary, and the hatchling turtles being a salmonella risk to children. It occurs naturally from Virginia to Iowa and south to Georgia and Texas—basically the

Red-eared Slider *Bred commercially in large numbers for the North American pet trade, the sale of baby red-eared sliders was banned some years ago because of the risk of salmonella, so they are now shipped in large numbers to Asia instead. It hibernates from October to March virtually buried in the pond-bottom mud.*
Photo: Clive Roots

Mississippi River Basin—and populations elsewhere such as in California and Maryland are believed to be due to human agency. Released or escaped animals have also colonized Japan, Australia, and several European countries. The red-eared slider prefers the still waters of ponds and lakes with muddy bottoms and dense water weed, and at the northern end of its natural range, in Iowa and Illinois, it hibernates from October to April, buried in the mud. It is a very wary species and for quick access to the water it basks on partially submerged logs and on mats of floating vegetation rather than on the pond banks.

Common Snapping Turtle (*Chelydra serpentina*)

This large turtle is a native of the lakes and rivers of North America, from southern Canada to Mexico east of the Rockies; and also occurs in Central America and northern South America. It is an aquatic species that basks infrequently. The "snapper" is Canada's largest freshwater turtle, an ugly and bad-tempered animal, which has a shell length of 18 inches (46 cm) and weighs up to 77 pounds (35 kg). Its shell is greenish-gray, but often appears green due to the algae growth; it seems too small for its body, which bulges out between the plastron and the small, rough carapace, which has three longitudinal keels. The snapper has a long tail and a large head and tremendously powerful jaws capable of inflicting a serious bite. Consequently, together with its large adult size and aggressiveness to other species and people, it has never been a popular animal for garden pools. It is a highly carnivorous, mostly bottom-feeding turtle, which also takes ducklings and muskrats from the surface. Although totally aquatic it can move well on land with its powerful limbs and claws. The snapper hibernates from late October to late April or even May in northern waters, but is active all year in the south. It burrows into the mud or may use muskrat lodges, and often hibernates in large congregations.

Spotted Turtle (*Clemmys guttata*)

The spotted turtle is an animal of the shallow and still waters of the marshes, ditches, and bogs of eastern North America from Quebec and southern Ontario to Florida. When alarmed during basking it usually moves slowly into the water, and if further threatened buries itself into the muddy bottom or hides among the debris. It is a small and attractive, dark greenish-brown turtle, seldom more than 4 inches (10 cm) long, with a yellow spot on each large scute of its carapace when juvenile, but increasing to several spots on each as it matures. However, the density of spots varies considerably, with some animals being very heavily spotted and others having very few spots on the carapace but then usually having spotted heads and necks. The spotted turtle hibernates underwater, buried in the mud, in the north of its range from October to April (see color insert).

Stinkpot (*Sternotherus odoratus*)

The stinkpot is a musk turtle, one of the commonest members of the family *Kinosternidae*, with a wide range across eastern North America from southern Canada to the Gulf of Mexico. It has glands under the carapace, the secretions from which have a very strong musky odor, and tiny barbels on its chin and throat, which distinguish it from other species in the genus. Males have longer and thicker tails with a horny nail on the end. Its shell is smooth and highly domed and is usually dark brown decorated with black spots or streaks, and with yellowish stripes on the sides of the head, one above and below the eye. The stinkpot is a highly aquatic turtle that spends most of its time on the pond bottom, but leaves the water to lay its eggs and to bask, even climbing sloping branches to limbs several feet high. Although a tiny turtle, just 5 inches (13 cm) long, it can be quite aggressive and is able to reach back to its hindlimbs to bite. Despite being highly aquatic, the stinkpot is a poor swimmer and usually walks on the bottom of its pond in search of food. It prefers the quiet shallow waters of ponds, marshes, swamps, sloughs, and canals, where it hibernates in the water under debris, buried in the mud, or in muskrat burrows.

European Pond Terrapin (*Emys orbicularis*)

This is a carnivorous terrapin of the ponds, ditches, marshes, and rivers with slow-moving water and lots of vegetation, occurring in Europe, North Africa, and Asia Minor, east to Iran and north into Azerbaijan and Georgia. It has an oval-shaped carapace that reaches a length of up to 12 inches (30 cm), and although there is considerable variation in color it is usually dark brown with yellowish spots and streaks. It has a long tail and a semiflexible plastron with a hinge that allows the front part to move up and down. The European pond terrapin is perfect for a garden pond in zones with mild winters, and breeds readily if it has access to a sandbank. It uses its tail to assist in digging a nest pit in which it lays up to fifteen eggs. It is now very rare and localized in Europe, having been affected by river and marsh drainage and by the introduction and establishment of the red-eared slider. At the northern extent of its range—in Germany and Poland—it hibernates each winter for several months in the mud at the bottom of its pond, while in the south—in Spain and Morocco—it rarely needs to hibernate but may estivate when its pond dries out. With a long incubation period (at least three months even in a good summer) its eggs may not hatch until September, and in the north of its range its hatchlings, like those of the painted turtle, may hibernate where they hatch and may therefore be similarly freeze tolerant. The adults hibernate in the mud at the bottom of their pond.

Reeves Terrapin (*Chinemys reevesi*)

This turtle is a favorite food animal in China and is consequently now very scarce there; it has also been a popular pet and reptile hobby trade species, especially in Europe and Britain. It reaches a length of 8 inches (20 cm) and has a dark-brown carapace and yellowish plastron spotted with brown, plus yellowish stripes on the head and neck. Its soft parts (the areas of skin between the shells) are usually dark also, either dark olive-green or black, and they often have solid or broken yellow lines on the neck. The carapace has three well-defined keels from front to back, and males have thicker tails than the females. The juveniles are greenish but darken with age. Reeves turtle is quite aquatic and hibernates underwater, buried in the mud, from October to April, beginning torpor when the water temperature drops below 53.6°F (12°C). It is a native of central-eastern Asia (China, Korea, and Japan) and is therefore one of the most northern-ranging turtles in that part of the world. It is omnivorous, and eats soft plants, aquatic insects, fish, tadpoles and frogs, and scavenges on carrion.

Spiny Softshell (*Trionyx spiniferus*)

The soft-shelled turtles are the quickest of all turtles; they swim fast and despite being highly aquatic can also move fast on land. Their hind feet are partially webbed, and their shells are flat, soft, and leathery as they lack the horny plates of other species and are therefore quite flexible. The spiny softshell is a large turtle, in which females reach 18 inches (46 cm) in length, but the males are usually no more than half the size. Olive-gray to yellowish-brown in color, the males have dark spots on their carapace plus tiny spiny projections that give the shell a sandpaper-like texture, and they generally have longer and thicker tails than females. The females have large blotches on the carapace and their shells are smooth except for the front, which is edged with spines. Its preferred habitat are rivers and lakes with sandy or muddy bottoms, and it is distributed throughout eastern North America from southern Canada to the Gulf of Mexico and the Mississippi River basin.

The spiny softshell has a long and flexible snout with nostrils at the tip, which allows it to breathe in shallow water with just the snout above the surface while the rest of the body is submerged. It lies buried in the mud with just its head exposed to ambush frogs, tadpoles, snails, fish, and crayfish, but it can also retract its head beneath its shell and then snap it out to capture prey with its powerful jaws. Softshells can remain submerged for several hours, drawing water into the pharynx to extract the dissolved oxygen through gill-like vascular papillae in the throat. They hibernate in the north from October to March, but not at all in warmer southern waters.

Wood Turtle (*Clemmys insculpta*)

This turtle occurs in the northeastern United States and neighboring Canada, and is a semiaquatic species of forested streams and rivers, which often leaves the water to wander some distance through woodland and fields, and is North America's most terrestrial turtle after the gopher tortoises and the box turtles. It feeds in water and on land, eating aquatic insects, snails, mussels, carrion, berries, and herbs. It also hibernates on land in communal burrows in riverbanks, from October to early April. It is just 7 inches (18 cm) long and is very rough-shelled, each of its large scutes rising up in an irregular pyramidal shape that is grooved and ridged. These resemble a cross section of a tree branch showing growth rings radiating from the center, and when the shell is dry and the grooves that run cross-wise over the rings are more obvious, it has the appearance of a spider's web. The wood turtle's carapace is dark brown, its plastron is yellow, and it has a black head and brown limbs with red or yellow on the throat and limbs. It is now very rare in many parts of its range due to commercial harvesting in the mid-twentieth century for the pet trade and biological supply houses, which completely wiped out some populations. It has a very long lifespan, with individuals often living for over fifty years.

Tortoises

The tortoises are terrestrial turtles that are characterized by their stumpy legs and blunt, heavily scaled feet. Their carapaces are hard-shelled and usually high-domed, an exception being the pancake tortoise (*Malacochersus tornieri*) which has a flattened shell. They are not as cold-hardy as the aquatic turtles and do not range as far north, occurring only in warmer latitudes, mostly in the tropics and subtropics of all the continents except Australia. They also occur in temperate regions experiencing very hot summers such as the American Southwest and the lands bordering the Mediterranean Sea both in Europe and Africa, where they hibernate for the inclement months. Although some have evolved in the tropical rain forests, tortoises are mostly animals of the hot and arid regions, including dry grasslands, rocky and sandy deserts, and scrubland. They have evolved various strategies to protect themselves from intense heat and dryness. The thickness of their shells helps to prevent desiccation, but they also burrow to escape the sun, and may estivate during the hottest part of the year. The desert tortoise (*Gopherus agassizi*) of the American Southwest stores water in its enlarged bladder, and the spurred tortoise (*Geochelone sulcata*) of the southern Sahara cools its neck and legs with its own saliva to prevent overheating. Unlike the aquatic turtles the tortoises are mainly herbivorous, but they do eat carrion and invertebrates when the opportunity arises. They are generally active at dusk and dawn and rest during the heat of the day. They have always been popular as pets, especially those from the Mediterranean regions and from the southern United States. Most of the species commonly kept as pets hibernate in the wild.

In addition to the tortoises the animals known as box turtles are also terrestrial. They are an endemic North American group which, unlike the tortoises, range well into the colder regions of the temperate zone—almost reaching the Canadian border in Michigan and Illinois. They are therefore the second-northern-most terrestrial turtles, and can tolerate freezing for a few days.

Greek Tortoise (*Testudo graeca*)

This species occurs in southern Europe, Asia Minor eastward into Iran and north to the southern Caucasus, and in North Africa. It is often also called the spur-thighed tortoise, but despite its name it is not widely distributed in Greece. Unfortunately, it is no longer common anywhere as it and others from the region were decimated by the pet trade, with annual imports into the British Isles during the middle of the twentieth century totalling 300,000 per year. If placed end to end, the specimens that arrived tightly packed in large wicker baskets during a five-year period in the 1960s would have stretched almost 170 miles (273 km), and the species was exterminated in many localities. The Greek tortoise reaches a length of 10 inches (25 cm) and is often confused with Hermann's tortoise, but its oval-shaped carapace is generally paler, being yellowish or olive-brown with black spots. It has a single plate at the back of the carapace above the tail, and it has a prominent spur on the back of each thigh, which are lacking in Hermann's tortoise. Scrubland and semiarid regions are this tortoise's preferred habitat, and it hibernates in the northern parts of its range for several months each winter, generally

Koi Carp   *Making a spectacular addition to an ornamental pond, koi carp are domesticated descendants of the wild carp that live naturally in eastern Asia but have been introduced into many countries. They are very hardy fish that hibernate when the water temperature drops to 46.4° (8°C), and can remain torpid on the pond bottom beneath the ice for several months.*

Photo: Brian Tan, Shutterstock.com

Collared Lizard   *A relative of the large green iguana, the collared lizard lives in western North America from Idaho south to Baja California and east to Missouri. In the north of its range it hibernates from October to March, while in Death Valley its winter dormancy is shorter but it also estivates in midsummer.*

Photo: Van Truan, Shutterstock.com

European Tree Frog *Some tree frogs are actually terrestrial animals, but this one is a true tree-climber that spends most of its time high in trees and finds all its invertebrate food there. Like all tree frogs it is dependent upon water in which to lay its eggs. With a drier skin than the typical frogs it sunbathes to warm up to its active temperature, and may darken its skin to increase heat absorbtion.*
Photo: Alexander M. Omelko, Shutterstock.com

Northern Leopard Frog *A cold-tolerant aquatic frog of the northern United States and Canada as far north as Great Slave Lake and Labrador, the leopard frog hibernates in the pond-bottom mud. In the far north it is active only from mid-May to September when its pond is ice-free.*
Photo: Bruce MacQueen, Shutterstock.com

Fire Salamander *The contrasting colors and pattern of this amphibian warn potential predators that it is a dangerous creature; the toxic contents of its parotoid glands has killed dogs that ate them. A cold-hardy species, in central Europe it hibernates only when the temperature reaches the freezing point, while in the southern extremity of its range (in North Africa) it may not hibernate, but estivates in midsummer.*
Photo: Carsten Reiner, Shutterstock.com

Red Eft *The red eft is a juvenile red-spotted newt that began life as an aquatic larvae, but then spends three years on land, never entering water and hibernating under loose soil and leaves. It will return to the water when it becomes an adult, and will hibernate there, but if its pond dries out it assumes a terrestrial existence once again and resembles a salamander.*
Photo: Michael Litvinov, Dreamstime.com

Broad-banded Copperhead *Copperheads are pit vipers, members of the family* Crotalidae *and therefore related to the rattlesnakes and moccasins, all highly venomous and dangerous reptiles. They have sensing pits below their eyes that can detect heat from nearby warm-blooded prey.*
Photo: Courtesy USFWS, R. Rauch

Gray-banded Kingsnake *A very variable snake in both color and pattern, and primarily nocturnal, this race of the kingsnake hibernates from mid-October to late March in Texas, and is then likely to estivate during hot and dry conditions in June.*
Photo: Courtesy Patrick Jean, Nantes Museum

Gila Monster   *The only venomous lizard in the United States, the Gila monster spends much of its life underground, avoiding the extreme temperatures of its southwestern desert habitat. It hibernates all winter and estivates during the hottest days of summer, being active in spring and then again in the late summer and fall.*

Photo: Courtesy Indianapolis Zoo

Spotted Turtle   A freshwater species from eastern North America, the spotted turtle buries itself in the pond bottom mud or debris when threatened. It also hibernates there from October to April in the north of its range (Quebec and New England), but usually finds hibernation unnecessary in Florida.

Photo: Photos.com

Painted Box Turtle   Box turtles are an American group that are considered intermediate in form and behavior between the aquatic turtles and terrapins and the terrestrial tortoises, and have a terrapin-like head and a typical tortoise high-domed carapace. But they are mainly terrestrial animals that rarely enter water and hibernate on land, in Michigan and Illinois between October and April, burrowing into loose soil and leaves.

Photo: Donald Rose, Shutterstock.com

Ruby-throated Hummingbird   *The supreme exponents of short-term or daily torpidity to conserve energy, hummingbirds become hypothermic when roosting if the ambient temperature drops below 95°F (35°C), which is just about everywhere, even in the tropics. With their small size and incessant activity they must feed continuously to fuel their high metabolic rate.*

Photo: Courtesy USFWS, Dean W. Biggins

Poorwill   *The only bird known to hibernate, the poorwill is an insectivorous nocturnal species of western North America. It becomes completely torpid for up to 3 months, its body temperature dropping to 41°F (6°C), during which time it survives on stored body fat.*

Photo: Courtesy Brian Currie

Gold-mantled Ground Squirrel   *A small ground squirrel often mistaken for a chipmunk but lacking their facial stripes, this rodent is a common resident of the mountain ranges of western North America. It seems reluctant to begin its long winter sleep and may not enter hibernation until the temperature is well below freezing. It awakens regularly all winter to feed from its underground stores.*

Photo: Photos.com

Yellow-bellied Marmot   *One of the more colonial species, this marmot hibernates in family groups, all cuddled together to reduce heat loss. It has a short and divided active season, generally hibernating from early September to March and then estivating in June.*

Photo: Courtesy National Park Service

from October to March, depending upon the locality. In Morocco it estivates during times of extreme heat and dryness.

Hermann's Tortoise (*Testudo hermanni*)

With a more restricted distribution than the former species, this tortoise lives in southern Europe, including southern France, Italy, the Balkans, and southeastern Romania. It is also smaller, with a carapace length of only 8 inches (20 cm), and its top shell is highly domed and generally appears much darker than the Greek tortoise; but its most characteristic feature is the large scale on the tip of its tail, which is lacking in *Testudo graeca*. It also has two supracaudal (over the tail) scales whereas the Greek tortoise has only one. The main external difference between the sexes is the length of their tails, the male's being longer. Hermann's tortoise lives on mainly dry sandy and scrubby hillsides and in evergreen oak forests, but also ventures into cultivated land and meadows. The Tortoise Village at Gonfaron near Cannes was established in 1988 to rehabilitate and breed this species, and now houses over 2,000 specimens, plus several hundred Greek tortoises. Many have been returned to safe areas or used to restock the Isle de Levant off the French coast, where they have military protection. This species hibernates

Hermann's Tortoise   *A once common terrestrial species (hence called tortoise in Europe) of several countries bordering the northern shores of the Mediterranean, Hermann's tortoise was over-collected for the pet trade in the mid-twentieth century and is now rare. It hibernates between November and March and may estivate during exceptionally hot and dry summers.*
Photo: Courtesy Paolo Mazzei

from November to March depending on the local climate, and may estivate if summers are exceptionally hot and dry.

Horsfield's Tortoise (*Testudo horsfieldi*)

A pale-golden tortoise with black markings, about 10 inches (25 cm) long, Horsfield's tortoise has an oval carapace that is flattened along the sides; it has a long neck, pointed tail, and tiny spurs. It varies in color from light tan to yellowish-green and has brown or black markings on the large scutes. It cannot tolerate dampness and lives on dry and barren rocky hillsides and sandy steppes, usually in the vicinity of springs or streams where vegetation is available. It occurs across most of Central Asia from western China through Uzbekistan, Kazakhstan, Pakistan, and Afghanistan to Iran, where its range extends north of the Caspian Sea to 52°N, the northern-most distribution of all the truly terrestrial tortoises. This region of the world is subjected to freezing winters and searingly hot summers, and in parts of its range Horsfield's tortoise hibernates from October to early April, then estivates in July and August, for a total of almost nine months each year. It is a burrower and its long periods of dormancy are spent underground, the microclimate several feet below the surface providing a suitable environment irrespective of the harsh conditions above. This tortoise's burrows are usually about 6 feet (1.8 m) long, and end in a sleeping chamber large enough for several to occupy; they are dug in spring when rains have moistened the hard soil. Its eggs hatch in September and the hatchlings may remain in the nest until the following spring.

Spurred Tortoise (*Geochelone sulcata*)

The largest land tortoise in Africa, and the third largest in the world after the giant tortoises of the Galapagos and Aldabra Islands, the spurred tortoise may have a carapace length of 31 inches (80 cm) and can weigh up to 220 pounds (100 kg). It lives in the Sahel, the dry grassland and semidesert region of frequent droughts and increasing desertification bordering the southern edge of the Sahara from Mauritania on the Atlantic Coast across the continent to Eritrea on the Red Sea. Its name is derived from the spurs on its thighs, and although it has a thick skin that provides protection from dehydration, it digs deep burrows down to moist sand to shelter in, by day or night depending upon the temperature, and it is often active at night when it is cooler. In times of extreme heat and drought it estivates in its burrows, which several tortoises may share and which are usually about 3 feet (90 cm) deep and 10 feet (3 m) long. The burrows also provide shelter for other Sahel animals, especially beetles, scorpions, and snakes. The spurred tortoise's carapace is brownish-yellow with dark-brown borders, and its skin is golden-yellow or light tan. The incubation period for its eggs, which are buried in the sand, is about seven months, and their hatching is timed to coincide with the period of scanty rainfall which should occur from June to September.

Desert Tortoise (*Gopherus agassizi*)

This is the largest indigenous tortoise in North America, one of three gopher tortoises that are the only members of the family *Testudinidae* native to the United States. Individuals regularly grow to 14 inches (35 cm) long and weigh about 15 pounds (6.8 kg). The desert tortoise's natural habitat is sandy and rocky desert and the mesquite scrub and chaparral of the Mohave and Sonoran deserts of Southern California, Nevada, Arizona, and neighboring Mexico. In these harsh regions it burrows not only to hibernate but also to estivate, and may spend up to 95 percent of its life in the burrow, protected from freezing temperatures from November to March and then from the hottest days of summer, when ground temperatures may reach 145°F (62°C). It has a high-domed carapace and elephant-like columnar hind legs, while the forelimbs are more flattened and muscular. It uses its forelimbs for digging burrows, but the female uses her hind legs for digging her nest. The desert tortoise has a gular horn—an extension of the front of the plastron—which is used in fighting, when an attempt is made to hook it under the opponent's carapace to flip it over. It rarely has access to standing water and has been known to survive without drinking for a year, obtaining sufficient liquids from its plant diet, which includes prickly pears. When water is available it drinks copiously and its large bladder can store 30 percent of the tortoise's weight in water and excretable wastes. To help conserve water the urates are separated and eliminated as semi-solids.

Eastern Box Turtle (*Terrapene c. carolina*)

Box turtles are an intermediate group, endemic to North America, which lie between the aquatic turtles and the land tortoises, but they are mainly terrestrial animals despite their name; they seldom have access to standing water, and then may only enter it to soak for brief periods. They are small animals, reaching a maximum of 6 inches (15 cm) in length, and have a broad hinge across the plastron about one-third back from the front, which allows the two sections to close tightly after their head, tail, and legs have been withdrawn. With this protection they are perfectly suited to life on land yet are in fact more closely related to the aquatic turtles than the terrestrial tortoises. Their high-domed carapace is reminiscent of the tortoises, yet they have terrapin-like heads (see color insert).

Although box turtles live farther north than the terrestrial tortoises, their range does not extend as far north as the aquatic species, and the highest latitude is reached by the eastern box turtle, which lives in the eastern United States from New Hampshire, Michigan, and Illinois south to Tennessee and Georgia. It is a small turtle, generally no more than 5 inches (13 cm) long, with a very variable carapace with yellow or orange blotches on a black or dark-brown background. It is omnivorous and eats plants, berries, insects, worms, and carrion. Along the mid-Atlantic coast box turtles may remain semiactive all winter, but northern populations must evade the cold and hibernate on land from late October to early April, usually under a thin covering of soil and leaves, although they may dig deeper. They are therefore quite dependent on a thick blanket of snow for protection, and in hard winters with poor snow cover are exposed to freezing temperatures. However, they are partially freeze tolerant, and can survive the freezing of almost 60 percent of their body water to temperatures down to 24°F (−3.6°C) for several days without injury.

Notes

1. There are also numerous species of lizards that are legless and superficially resemble snakes, except for having eyelids and ear openings, which snakes lack.

2. Which practice ectothermy—temperature regulation that depends upon the absorbtion and dissipation of heat from and to their environment.

3. Whose body temperature depends mainly upon heat produced by metabolism and dissipation of the excess to the environment.

4. Some mammals, for example, the sloths and monotremes (the platypus and echidnas), have very low body temperatures.

5. The green anole is an exception, as it does not hibernate in Tennessee.

6. Seabirds that dig their nest burrows on the islands, and whose young are known as muttonbirds.

7. Some iguanid and monitor lizards were recently found to have venom-producing glands.

8. Scale-less snakes have occurred occasionally in several species as a result of a genetic defect. They have a coarse, rubbery appearance.

9. Painted turtles living at higher latitudes, such as in Alberta's Peace River region, are believed to have originated from introductions.

5 Regulated Hypothermia

There are two forms of sleep involving lowered metabolism in warm-blooded animals—long-term hypothermia or hibernation (including estivation) and short-term hypothermia or daily torpor, and both are regulated by the sleeping animal. With their ability to migrate to a more favorable environment for the winter, hibernation is a very poorly developed characteristic in birds. In fact, only one species is known to hibernate, although it is suspected in others. Birds migrate south for the winter mainly because their source of food is no longer available, not necessarily because they cannot withstand the cold. This is particularly so for the soft-billed birds—those that eat insects, fruit, or nectar—all of which are in short supply in temperate regions in winter. Some birds that glean insects from tree trunks or within the tree itself such as nuthatches and woodpeckers, or that are mainly berry-eaters like the waxwings, can stay in cold temperate regions all winter, but most insectivorous species must migrate to eat. It is therefore rather surprising that the common poorwill (*Phalaenoptilus nuttallii*), the only bird known to hibernate, is purely insectivorous, and instead of migrating to the tropics it spends the winter in a state of reduced animation in its summer range.

In the eighteenth century, when swallows vanished for the winter it was believed they had buried themselves in the mud or were hiding in tree holes or crevices until spring. The discrediting of this myth, when their migratory habits and real winter whereabouts became known, resulted in scant attention being paid years later when it was suggested that perhaps some birds did hibernate after all. During his studies of the white-throated swift (*Aeronautes saxatalis*) on Stover Mountain in California in 1913, E. C. Hanna found several birds in a numbed condition in a crevice. As the swifts did not appear on cold days, he concluded that they had an intermittent hibernation, but his reports did not attract much attention. This changed in December 1946, when Edmund Jaeger found a poorwill in a profound state of torpor in the Chuckwalla Mountains of California's Colorado

Desert. The bird's respiration and heart rate were so low they could not be detected with the equipment Jaeger had with him. Later readings of its temperature on several occasions averaged 64.5°F (18°C), far below the 105°F (40.5°C) considered normal for the species, while the air temperature varied between 62.6°F (17°C) and 75.2°F (24°C). Since then the lowest temperature that has been recorded for a hibernating poorwill is 41°F (6°C).

The poorwill was hibernating in a niche in a large boulder, protected from vertically falling rain or snow by the depression, but not from angled storms, and on one violent night its tail was frayed by high winds and heavy sleet; yet it was totally oblivious to its surroundings and happenings, and even shining a flashlight into its eyes brought no response. This bird was banded and monitored for three years and it returned to the same site each winter, hibernating for periods of up to eighty-six days. It has since been determined that ⅓ ounce (10 g) of fat would sustain a torpid poorwill for 100 days.

Daily torpor or short-term hypothermia to conserve energy is a far more common phenomenon in birds than hibernation, and is known to occur in many species. Some become lethargic in times of stress—mainly low temperatures and shortages of food—but a few actually become hypothermic every day of their lives irrespective of the temperature or food availability, which results in considerable energy savings.

Birds are warm-blooded or homeothermic, able to regulate their temperature within a narrow "set range," and with a high level of basal metabolism compared to the poikilotherms or cold-blooded animals. Their normal temperature range is 99.9°F to 110.3°F (37.7°C to 43.5°C), with most species between 104°F (40°C) and 108°F (42°C). Those that can lower their temperature are said to be heterothermic—having the ability to temporarily enter a lower state. They include a number of very diverse species, with considerable variation in their distribution, habitat, behavior, and physiology, and include nectar-eaters, seed- and grain-eaters, insectivorous species, carnivores, and even some that eat carrion.

Daily hypothermia is the condition in which the core temperature falls below normal, producing a state of torpor or lethargy, usually as an adaptive response to unfavorable environmental conditions. It is caused by self-induced reduction of the metabolic rate and body temperature, although they rarely decrease as much as in regular hibernation, and thus the bird's demands for energy. This torpidity can occur day or night, throughout the year or just seasonally, and is mainly a condition of the sleeping diurnal bird at night, simply because there are more day-active species than nocturnal ones. But it also occurs in reverse, lowering the body temperature of nocturnal birds such as the frogmouths and nighthawks when they rest during the day.

Recent studies and research prove that regulated hypothermia is more widespread in birds than previously believed, and it is known to occur in eight orders of birds, the *Caprimulgiformes* (nighthawks and frogmouths), *Apodiformes* (hummingbirds and swifts), *Coliformes* (colies), *Columbiformes* (pigeons), *Cuculiformes* (cuckoos, anis, and roadrunner), *Falconiformes* (falcons), *Ciconiiformes* (vultures), *Strigiformes* (owls), and various members of the *Passeriformes*, including sunbirds, manakins, chickadees, white eyes, swallows, finches, and the smooth-billed ani. While it has great survival value in terms of conserving energy and combating the

Common Nighthawk *Rarely seen in such an exposed position in daylight the common nighthawk is a nocturnal aerial insect-hunter that is more often heard than seen even at night, when it dives and then zooms up quickly with whirring wings. It is one of an increasing number of birds known to become hypothermic when resting by day, to save energy, but it does not hibernate like its relative the poorwill, preferring to spend the winter in South America.*
Photo: Courtesy National Resources Conservation Service

elements, it does increase the risk of predation, as a lethargic bird cannot react as quickly to a threat.

Most birds of the cold temperate zone migrate south for the winter, but a few are permanent residents, nesting there in summer and surviving the low temperatures in winter, in company with others that may enter the region from even farther north where they have spent the summer. Winter torpor is therefore a natural condition in many of these birds, especially small species below 2.8 ounces (80 g) in weight, which can substantially reduce their body temperature to conserve energy. But daily torpor is not just a winter phenomenon, for even in northern summers some birds lower their temperature nightly irrespective of the conditions, and others do it during cold spells or when food is short. Many tropical birds also become hypothermic at night, even though the ambient temperature may be only a few degrees cooler than daytime levels.

There are two forms of short-term lethargy in birds. The first, in which the temperature drops only a few degrees, usually by about 9°F (5°C) but in some species down almost by 18°F (10°C) below normal, occurs in a number of birds of variable size, such as cuckoos, pigeons, vultures, some birds of prey, the roadrunner, and several passerine birds. It has been given various names including moderate hypothermia, rest-phase hypothermia, and facultative hypothermia.

As an adaptation for cold desert nights roadrunners may become hypothermic, lowering their core temperature by 10.8°F (6°C). They love to sunbathe, and in the early morning position their scapular feathers (those covering their shoulders) so that the black skin beneath is exposed and absorbs heat, which helps them to quickly regain their normal body temperature while reducing the use of their own energy stores.

Several birds of prey are also known to practice daily hypothermia, and it probably occurs in many more. It is a valuable adaptation for conservation for these birds, especially vultures, which have an unpredictable food supply. To save energy both the black vulture and turkey vulture lower their temperature from 100.4°F (38°C) to 91.4°F (33°C) when roosting, and red-tailed hawks become hypothermic when they are inactive or food is short, reducing their body temperature by 12.6°F (7°C) below normal.

Daily torpidity also occurs in several pigeons and is likely to be even more widespread among the pigeons and doves. When food is short diamond doves and Namaqua doves can lower their body temperature by up to 12.6°F (7°C) at night when they are inactive, reducing their energy consumption by 10 percent. When they were starved domestic pigeons gradually lost weight, as expected, but then they became hypothermic at night, with lowered body temperature and metabolic rate, to compensate for their starvation.

The tubenoses (*Proceleariiformes*), which include the petrels and storm petrels, are highly pelagic birds of the world's oceans. Some are migratory and spend their summers in the rich polar seas, returning to the tropical latitudes for the winter. The parent birds normally forage all day on the open ocean, but when they have difficulty finding sufficient food they do not return to the nest and eggs, and their embryonic chicks (in the egg) have survived suspended incubation for three days. If the same food shortage situation arises after the eggs have hatched the chicks become torpid while awaiting their next meal.

The other form of daily hypothermia is a much deeper one, involving greater cooling with a temperature drop of more than 18°F (10°C), and in some cases considerably more. Associated with this is a reduced heart rate and lowered respiration, sometimes down to one or two per minute from several hundred. These are all attributes of full hibernation, but they occur only for short periods, generally overnight but in some species for several days at a time. Birds that practice this deep or pronounced torpor and its concurrent large savings of energy are the hummingbirds, some swifts, the common poorwill, and the tawny frogmouth.

The hummingbirds, with their very active lifestyle, small body size, and large food requirements to power their rapid metabolic rate and high body temperature, conserve energy by lowering their temperature every night when the air temperature is below 95°F (35°C), and they can cool down to 53.6°F (12°C). The body temperature of the nighthawks and swifts drops frequently to 68°F (20°C) and has matched even lower ambient temperatures. In the mountains of New South Wales a tawny frogmouth's temperature dropped from its normal 104°F (40°C) to 83°F (28.4°C) overnight when the air temperature was 12.6°F (7°C).

There have also been many accounts in the bird literature of swifts becoming torpid when deprived of food during cold weather. This includes the white-throated swift (*Aeronautes sxatalis*), the common swift (*Apus apus*), Vaux's swift (*Chaetura vauxi*), and the chimney swift (*Chaetura pelagica*), in which even full hibernation has been considered a possibility. Nestling swifts also become torpid when their parents cannot find sufficient food for them during unseasonably cold weather, and during several days of fasting, which would be fatal to many birds, they draw on their fat deposits and tissues to maintain their metabolism, but may still lose 50 percent of their body weight. The normal temperature of nestling swifts is 99°F (37°C) to 102°F (39°C), but when they were starving it fell to 70°F (21°C) at night but returned to normal during the day.

Hypothermic-prone birds do not usually become torpid when they are incubating their eggs as they would be unable to provide the correct incubation temperature, which averages 104.9°F (40.5°C). However, some hummingbirds do continue their regular nightly torpor while nesting, and their eggs have hatched after being cooled to 45.5°F (6.5°C) overnight. Eggs that contain well-developed embryos are assumed to have greater survival value when left unattended, due to the slower loss of heat, than eggs in the very early stages of incubation.

Birds save considerable energy by becoming torpid rather than just maintaining their high body temperature when asleep. Hummingbirds, which lower their temperature by 18°F (10°C) at night, make savings of at least one-quarter of their normal overnight energy expenditure. However, any savings in energy must be offset by the cost of the energy needed to return the bird to its normal temperature. Arousal time from torpidity depends upon the temperature level of the torpid bird but in small species (except the hummingbird) it is usually quick and does not use a great deal of energy. In large birds the calorie savings through becoming torpid are relatively small as they must use considerable energy to warm up afterward unless they can sunbathe.

Although several birds, especially the larger species such as vultures and hawks, can practice daily hypothermia for several days without eating, the smaller species that become torpid daily, such as the hummingbirds and chickadees, must eat as soon as they awaken each day and replenish their reserves for the next night.

Some of the Species

Poorwill (*Phalaenoptilus nuttallii*)

This is North America's smallest nightjar, just 8 inches (20 cm) long and weighing 1½ ounces (42.5 g), with a short bill and very wide gape. It is cryptically colored, with gray and black mottled plumage above, paler underparts with dark mottling, and a gray tail with dark bars, yet with a conspicuous white band across the throat. It is named for its melancholy repetitious call, which sounds like "poor will," with the first note lower than the second. The poorwill lives in western North America from southern Canada to Mexico, preferring dry open country such as chaparral, scrub, dry grasslands, rocky canyons, and sagebrush. The northern birds move south to California and Mexico for the winter. It is not an aerial hawker like the swifts, and catches its insects—moths, beetles, and grasshoppers—on the ground or flies up to seize them in the air. It nests on the ground, although it does not actually make a nest, but simply lays its two eggs on gravel or rock. The poorwill is the only bird definitely known to hibernate, becoming completely torpid for almost three months with a body temperature as low as 41°F (6°C) and undetectable heartbeat and respiration. It can also become hypothermic for several days, as it has done in southern Canada during late summer snowstorms (see the color insert).

Tawny Frogmouth (*Podargus strigoides*)

This is the largest of the fourteen species of frogmouths and is also the largest nocturnal insectivorous nightbird. It is the size of a crow, with a heavy body, large head, and a wide gape similar to a frog's. It lives in Australia and New Guinea, in a range of habitat excluding the dense rain forest and deserts, and when asleep during the day it adopts a motionless posture on an exposed branch in a vertical position with its head pointed upward. Its cryptic plumage of mottled grays and browns blends so well with the surrounding branches and foliage that it is very difficult to see. The frogmouth's large mouth, which was originally believed to have evolved for scooping insects out of the air while in flight, is actually used for plucking insects off the leaves or from the ground. It also eats snails, frogs, lizards, small birds, and mammals.

Frogmouths studied in winter at an elevation of 3,280 feet (1,000 m) in the eucalyptus woodlands in New South Wales became torpid, their body temperature dropping from 104°F (40°C) to 93.2°F (34°C) on most nights, and on very cold nights when the air temperature was 44.6°F (7°C), they dropped to just below 83°F (28.4°C). Although nocturnal, and normally resting all day, in winter they changed their habits. After a brief period of activity in the evening when they searched for food they became torpid for the night, and aroused just before sunrise to hunt; however, they very soon became torpid again until evening. They have excellent night sight and good binocular vision as their eyes face forward like the owl's, and they also have large, rounded wings with soft feathers and silent flight.

Common Swift (*Apus apus*)

The master of the air, a superb flier, and the most aerial of all birds, the common swift is so adapted for flight that it never settles on the ground like the swallows or martins, with which it is so often confused. Except when nesting it

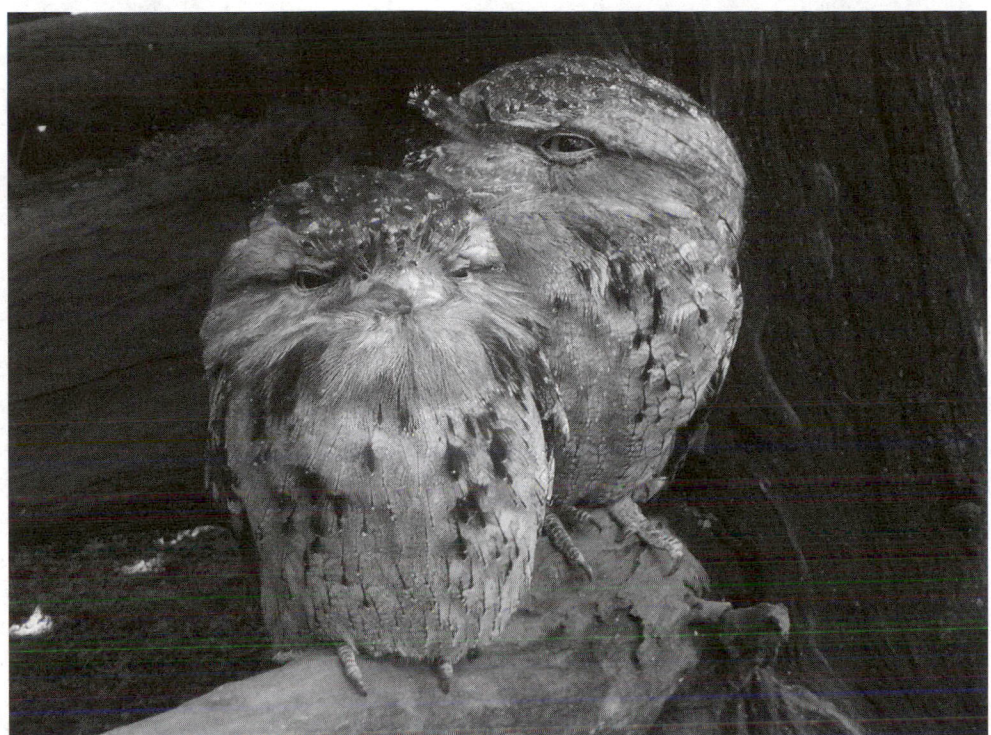

Tawny Frogmouths *Crow-sized, with large insect-gathering mouths, the nocturnal tawny frogmouths lower their metabolism when resting during the day. In winter in the Australian Alps their body temperature dropped to 83°F (28.4°C) on very cold nights, saving considerable energy at a time when food was scarce.*
Photo: Angela Davis, Dreamstime.com

spends its time in the air, sleeping and feeding on the wing, and even collecting its nesting material in flight, plucking windblown feathers, grass, leaves, and even pieces of paper out of the air and gluing them together with saliva. It builds its nest in a hollow tree, in a cave, or under the eaves of buildings like the swallows. It cannot perch, as it has such tiny feet, with four toes that all point forward, but may hang from a branch for short periods.

The common swift is dark brown except for its white chin; it is about 8 inches (20 cm) in length and has long and tapering scythe-shaped wings that span 17 inches (43 cm). It spends the summer months in Europe and Asia Minor and when insects become scarce it migrates south to central Africa for the winter. However, during cold spells in its summer range, and when food is scarce, it can enter deep torpor similar to the hummingbirds, with a drop in temperature from 105°F (40.5°C) to 55°F (12.2°C), and thus saves considerable energy. This dormancy is an extended form of daily hypothermia which may last for two or three days. In cold weather the parents may spend long periods on the nest huddled on top of each other, and on the chicks. Their nestlings are also able to survive cold weather and lack of food by becoming torpid at night, and when necessary for several nights until their reserves are depleted or their parents can hunt again. Whether their eggs are cold-hardy like those of some other birds is unclear, but on abnormally cold days swifts have been seen throwing all the eggs out of their nests.

Hummingbirds

The hummingbird family *Trochilidae* contains over 300 species of highly specialized birds which occur throughout the New World from Alaska to Tierra del Fuego, and from sea level to above the tree line in the Andes. They range in size from the world's smallest bird and therefore the smallest to practice daily torpor—the bee hummingbird of Cuba and the Isle of Pines—which measures just 2½ inches (6.3 cm) and weighs only 1/15 ounce (2 g), to the giant hummingbird which is 8 inches (20 cm) long and weighs ¾ ounce (21 g). The hummingbird's small size, high temperature, which averages 104°F (40°C), and incessant activity results in a high metabolic rate, with a respiration rate of up to 400 per minute and a heart rate that can reach 1,250 beats per minute. Even in the tropics hummingbirds become torpid at night to conserve energy, their breathing rate decreasing, their body temperature dropping to 48.2°F (9°C), and their heart rate down to 36; and they cannot fly the next morning until their body temperature reaches 95°F (35°C). Daytime torpidity has occurred in captive birds but is considered more likely due to malnutrition than lowered environmental temperatures.

Most experiments and observations have concluded that hummingbirds need at least half their weight daily in easily digested simple carbohydrates—sugar, honey, or nectar—but this alone is insufficient. They also need fat, protein, vitamins, and roughage from the regular intake of pollen and small insects. Their digestive rate is so rapid that even the remains of insects' chitinous shells are voided within ten minutes of being swallowed. The hummingbird's bottom beak is flexible and bends to allow it to open wide to catch insects on the wing.

Hummingbirds also store fat for their long migratory flights. The ruby-throated hummingbird's stored fat increases its bodyweight by almost 50 percent, and allows it to fly nonstop across the Gulf of Mexico, a distance of almost 1,000 miles (1,610 km) from the Louisiana coast to the Yucatan Peninsula. It has been calculated that 1/14 ounce (2 g) of fat would provide sufficient energy for a hummingbird to fly nonstop for twenty-four hours averaging 27 miles per hour (43 kph).

Birds entering torpor raise and bristle their feathers, lose heat, and become lethargic, and although this obviously wastes heat initially, it does result in an economy overnight; however, the amount of energy saved by torpor depends upon the air temperature. It has been determined that a hummingbird that lowers its temperature to 60°F (15.5°C) burns only about one-fifth of the fuel needed to maintain its body at its normal temperature. When held, a deeply torpid hummingbird moves weakly and it cannot regain its grip if returned to its perch, but the rate of recovery varies, and some tropical species make take an hour to return to normal activity. During arousal from torpor a hummingbird's body temperature returns to normal at the rate of 1°F (0.5°C) per minute.

In Brazil, the torpidity of captive hummingbirds has lasted up to fourteen hours in the cool season, and they only became active after the morning sun had burned off the chill. However, North American species come out of torpidity at dawn, even arousing before it is light, so there appears to be no clues from the environment to trigger their arousal, which must therefore be controlled by a circadian rhythm—a twenty-four-hour internal biological clock. Anna's hummingbirds are active at daybreak in spring on Vancouver Island when the temperature is only 35.6°F (2.0°C), so they must begin arousing while it is still dark.

The metabolism of most birds when resting decreases as the air temperature rises, because they expend less energy to keep warm. In contrast the torpid hummingbird's metabolism increases like that of a cold-blooded animal, until at 95°F (35°C) the rates equal out. Hummingbirds live at up to 15,750 feet (4,800 m) in the Andes, where the temperature drops below freezing every night. At such altitudes they must avoid open areas when roosting, where they would be exposed to snow or cold rain, or clear skies would allow radiation of the earth's warmth and their own body heat to escape; some species sleep in caves, mine shafts, rock crevices, and similar protected areas where the temperature is slightly above that of the outside air.

The hummingbird's energy use is the greatest of all birds in relation to its size, in fact of all warm-blooded animals, and there are two reasons for this. First, because the basal metabolic rate is related to mass, with small birds expending more energy per unit of weight than large ones. Their small size, which exposes more body surface for each gram of weight, loses heat rapidly and therefore uses more energy to maintain their high body temperature. Second, their rapid flight and hovering—which requires the wings to be powered for the upstroke as well as the downstroke—demands additional energy. When the air temperature is 75°F (24°C), a hovering hummingbird consumes oxygen at least seven times faster than when it is resting (see color insert).

Black-capped Chickadee (*Parus atricapillus*)

This chickadee is one of the most familiar winter residents in Alaska, Canada, and the northern United States, a tiny, agile, and active bird, and a regular visitor, usually in small flocks, to winter bird tables for peanuts and sunflower seed. It is easily recognized by its solid black cap and throat, which are separated by white, and its gray back and buff sides. It survives extremely cold winters by caching food, having dense winter plumage, and by roosting in a sheltered place, usually a tree cavity. Even so, it must store energy during each short winter day for the long winter night, and must feed immediately upon awakening the next morning. The chickadee has low fat reserves, which have been measured at just 7.5 percent of its body weight, yet after a short feeding day it must then spend perhaps sixteen hours roosting in subzero temperatures. To save energy, the chickadee becomes hypothermic, reducing its metabolism and thus its core temperature from its normal 107.6°F (42°C) to 89.6°F (32°C), resulting in energy saving of almost 25 percent, at an ambient temperature of 32°F (0°C). At lower temperatures, chickadees do not continue to reduce their temperature to save energy but maintain it at 86°F (30°C) by shivering.

Turkey Vulture (*Cathartes aura*) and Black Vulture (*Cathartes atrata*)

The turkey vulture is the common vulture of much of the United States and southern Canada, and is joined in the east by the black vulture, with both species then ranging down through Central and South America. The turkey vulture is a

Black Vulture   With an irregular and unpredicable supply of food, scavengers like the black vulture (and the other North American species the turkey vulture), practice daily hypothermia to save energy. Their body temperature drops about 9°F (5°C) when they are roosting at night.

Photo: Stephan Ekernas, Dreamstime.com

familiar bird with a characteristic flight profile, soaring and wheeling with wings held above the horizontal—forming a two-sided or dihedral angle in the form of a shallow V. Called "buzzard" in the southern United States, it has a blackish body and two-tone wings, the flight feathers being paler, and it has a body length of 30 inches (76 cm) and a wingspan of 6 feet (1.8 m). At close range its red head is unmistakable, and it has thin, pinkish legs, but immature birds have black heads and are frequently mistaken for black vultures. The turkey vulture breeds from southern Canada to the Falkland Islands, and winters as far north as New Jersey in the east, but western birds migrate south to coastal California, the Gulf Coast, and Mexico for the winter. The black vulture is slightly smaller and has a wingspan of only 5 feet (1.52 m). Up close its black head is unmistakable, and in flight it has a short and square tail, whereas the turkey vulture's is longer and slimmer. These birds eat virtually anything, animal or vegetable, including carrion, offal, and excrement, rotting fruit and vegetables, and may also take young waterbirds, such as ibis, egrets, and herons, from their nests. They are one of the few birds with a good sense of smell, and can detect rotting carcases in dense undergrowth. The body temperature of both species drops by about 9°F (5°C) overnight to conserve energy, and they shiver to warm up when they arouse.

Leach's Storm Petrel (*Oceanodroma leucorhoa*)

This small seabird is a "tubenose," one of the *Procellariiformes*, which are pelagic birds that spend most of their lives at sea. Unlike all other birds they have tubular nostrils—an extension of their nostrils in the form of an open-ended tube on either side of the upper bill. These tubes are used primarily to excrete salt, as the birds are forced to drink salt water during their months at sea, and without this ability the excess salt would lead to kidney failure and death. It is the smallest of the tubenoses, a little larger than a bluebird, and is black with a white rump and has a forked tail. It was called Mother Carey's chicken by ancient mariners, and the name petrel is believed to be derived from St. Peter, as the bird seemingly walks on water, just dipping its feet while fluttering over the surface plucking up plankton. It spends months at sea in the northern Atlantic and northern Pacific oceans, usually hundreds of miles from the nearest land, and visits islets in the northern oceans to breed, arriving in darkness and nesting in shallow burrows or under rocks and logs. On land it is completely nocturnal, and excavates its nest tunnel in the dark and then shelters in it during the day. On land its weak legs cannot support its body, and it shuffles along using its wings as crutches. Parent storm petrels normally forage all day on the open ocean, but when they have difficulty finding sufficient food their embryonic chicks in the egg can survive un-incubated for three days. If the same situation arises after the eggs have hatched the chicks in the burrow become torpid when the parents do not return regularly with food.

6 Deep Sleepers

Rodents are the most plentiful of all mammals in both species and numbers. Unfortunately, their name is synonymous with rats and mice and long naked tails, but there is such great diversity in their shape, size, color, and coat that early biologists could not believe they were related and had difficulty identifying and classifying them, until they checked their teeth. Wild rodents have a reputation for destruction and contamination and for spreading disease, as they or their parasites harbor bubonic plague, tularaemia, leptospirosis, Rocky Mountain spotted fever, and other highly infectious diseases. Despite this, domesticated rats, mice, and gerbils are common pet animals, and rodents like the golden hamster, which lacks the typical bare tail, are even more popular. Rodents are a very important link in the food chain. Most species, especially the smaller ones, have large and frequent litters and are the main source of food for many predators, from owls and snakes to small cats and dogs. Some are able to take advantage of good times to "explode" in numbers; this in turn benefits their predators, which also have a good breeding year.

Rodents are gnawing mammals, characterized by their specialized jaw muscles and dentition, with continually growing, large yellow or orange incisor teeth. These gnawing teeth have a thick layer of enamel on the outside only, which forms a chisel edge, with the opposition of the top and bottom teeth preventing over-growth, and their chewing teeth are also incompletely enamelled. Rodents lack canines and premolars so there is a gap known as the diastema between their cutting incisors and the chewing molars. Gnawing is accomplished by holding the top incisors against the object while the bottom jaw moves its incisors forward and upward against the object, meeting the upper incisors. The molars do not meet while this is happening, allowing the animal to gnaw without abrading its back teeth or swallowing the gnawed material. The reverse happens when a rodent chews, for the lower jaw moves backward slightly, opposing the molars but not the incisors, so it can chew without grinding down its chisel-teeth.

In mammalian taxonomy the classification[1] of the rodents and their grouping into three suborders is based upon the structure of their skulls, jaws, and their chewing or masseter muscles, but membership in a particular group does not necessarily mean close resemblance, and many are quite dissimilar externally. Rats and mice, for example, belong to the same suborder as the hamsters, and the kangaroo rats, which resemble miniature marsupials, are grouped with ground squirrels and chipmunks. There are no hibernators in the suborder *Hystricomorpha*, and the rodents that hibernate belong to the following suborders:

Myomorpha—the rat-like and mouse-like rodents, which include the dormice, hamsters, jerboas, birch mice, and jumping mice. Their lateral and deep masseter muscles thrust the lower jaw forward to provide the gnawing action. The deep masseter muscle passes from the lower jaw through the eye orbit to the muzzle.
Sciuromorpha—the squirrel-like rodent group, which includes the ground squirrels, marmots, chipmunks, and flying squirrels, and the pocket mice, kangaroo mice, and kangaroo rats. Their lateral masseter muscles extend onto the snout in front of the eyes, and move the jaw forward in gnawing. The shorter deep masseter muscle closes the jaw.

■ HIBERNATION AND ESTIVATION

Rodents are adapted to survive in some of the world's most inhospitable regions, where food is scarce in winter and the land searingly hot in summer. As they cannot migrate long distances they form the largest group of mammalian hibernators and estivators, and have been called true or profound hibernators as their temperature drops close to freezing, and in at least one species actually below freezing. When dormant their metabolic functions are reduced to the absolute minimum for sustaining life, as illustrated by the alpine marmot, one of the best examples of evolutionary modifications for survival in a very harsh winter environment. While in hibernation torpor it takes just two or three breaths per minute and its oxygen consumption decreases from 600 cc to 30 cc per kg of body weight; its temperature drops from 99°F (37.3°C) to about 41°F (5°C); and its heart beats only three or four times each minute instead of the usual 120 times.

With such perfect hibernation behavior the rodents have been able to colonize some of the world's harshest regions, where survival hinges upon avoiding extreme weather, both hot and cold. They live within the Arctic Circle where they search for plants on the open tundra during the short summer. They survive in alpine regions high in the mountain ranges in both the New and the Old Worlds, and they inhabit midcontinental grasslands that experience many degrees of frost and deserts that have the highest temperatures ever recorded. However, hibernation does not necessarily guarantee their total safety, for studies of ground squirrel populations before and after hibernation have shown that there is high mortality of both adults and juveniles during their long sleep. They are subjected to heavy predation and natural disasters such as lack of snow cover and deep penetration of frost down to their den level and spring flooding. Young animals may be unable to store sufficient food for the winter and then cannot combat low temperatures.

The rodent hibernators vary in size, from the hoary marmot which weighs 18 pounds (8 kg) to the smallest terrestrial (nonflying) hibernator—the birch mouse—which weighs just ⅕ ounce (5.6 g). Their lifestyle includes a very long sleep, up to almost nine months each year for the high northern species, followed by a short summer into which all the normal activities of life must be compressed. These include avoiding their predators; maintaining a territory, finding a mate, mating, giving birth, raising a family, and then finding sufficient food for daily functions as well as either for storing as fat within their bodies, or making a cache of seeds and nuts for winter use. They also have to prepare a den or nest to protect them during their hibernation.

While all the rodents included in the species accounts that follow are hibernators, many live in such extreme climatic zones that they are also forced to estivate during the hottest months of the year, and in a few species their summer sleep may continue directly into hibernation. When the air temperature rises above 98°F (36.7°C) in the Mohave Desert (usually in July and August) the Mohave ground squirrel (*Spermophilus mohavensis*) estivates and then goes straight into hibernation until February or March. In the Craters of the Moon National Monument, yellow-bellied marmots (*Marmota flaviventris*), Columbia ground squirrels (*Spermophilus columbianus*), and yellow pine chipmunks (*Eutamias amoenus*) estivate in midsummer when the air temperature is above 90°F (32.2°C) and the lava temperature 170°F (76°C), until cooler and moister weather returns. They then hibernate to avoid the subfreezing temperatures of midwinter and up to 24 inches (61 cm) of snow. A captive thirteen-lined ground squirrel (*Citellus tridecemlineatus*) estivated in July and August and was active for two weeks before going underground again and not reappearing until March 15. In Utah, Uinta ground squirrels (*Spermophilus armatus*) entered estivation and hibernation combined in July and did not reappear until April. On the hot and dry Turkestan steppes, where the vegetation withers in early summer, the large-toothed souslik (*Spermophilus fulvus*) commences its estivation in June, plugging its den entrance with grass to keep out the heat, and then continues straight into hibernation, remaining underground until March.

The major requisites for successful hibernation are the selection or preparation of the den, and the storage of food. No mammal can undertake a long sleep without the necessary food reserves, which may be stored either internally or externally. The sources of energy of these sleeping rodents are lipids from their own body stores laid down in advance in preparation for their dormancy, or stored foods that they eat when they awaken regularly. During their dormancy they use about 15 percent of the energy used during normal activity, although the actual amount depends upon the temperature of the den, and thus their body temperature.

Internal food storers that lay down a supply of fat within their bodies include dormice, flying squirrels, marmots, ground squirrels, prairie dogs, jumping mice, and birch mice. When their winter supplies are stored internally they are at least

Thirteen-lined Ground Squirrel *Sought by foxes, badgers, hawks, and several other predators the thirteen-lined ground squirrel is a wary and solitary animal. Unlike the colonial prairie dogs, it lacks the benefit of many eyes watching for danger. Dormant from late October to March in North Dakota, it is one of the few rodent hibernators that stores fat internally and may also have a larder of stored foods.*
Photo: Clive Roots

safe and cannot be stolen. The external food storers that stock larders, usually within their burrows, with a large amount of dried grass, seeds, nuts, and berries, include hamsters, chipmunks, some ground squirrels, kangaroo mice and rats, pocket mice, and some dormice. Storing food does not necessarily imply hibernation, however, and a typical example of a nonhibernator is the pika, which stores dried grass among the rocks for winter use but does not become torpid and is out daily eating from its hay pile. Stores of seed and nuts are a magnet to other animals, and even humans have raided the stockpiles stashed away by rodents.

Both kinds of storers sleep for long periods, but it is not uninterrupted sleep, for even the internal food storers awaken regularly (usually once weekly) for a few hours, move around, shiver, and raise their temperature. Arousals for such short periods during hibernation may be due to the need to defecate and urinate, and it has also been suggested that it may stimulate the immune system to repel pathogens, and possibly even to allow the animal to sleep in order to regenerate the brain cells, as hibernation is not true sleep since there is no brain activity. Hibernators are also awakened by their nervous system when the temperature drops too low, and shivering and activity then brings it up to its set point to prevent them from freezing.

The business of storing either food or fat is not mandatory for all species, however, and several hibernators store food both internally and externally. The thirteen-lined ground squirrel is one of these, laying down a layer of fat but also storing food to eat during its periods of wakefulness.

Richardson's ground squirrels behave somewhat differently, for they are known to store seeding grass heads in their nest chamber, but are believed to save these for when they awaken in spring, rather than consume them during their numerous periodical arousals during the winter when stored body fat supplies their energy. But what actually happens below ground all winter is still largely unknown for many hibernators.

■ LENGTH OF HIBERNATION

A rodent's hibernation period depends upon the conditions in its immediate locality, and the most influential factors are latitude and elevation. It also depends upon the effect of the oceans, and in the midcontinental prairies of North America and the steppes of central Asia, the climate differs hugely from coastal regions at the same latitude. The harsh winters there are reminiscent of the polar regions, requiring lengthy hibernation periods, yet the summers may be hot enough to induce estivation. For example, the black-tailed prairie dog hibernates for five months on the Canadian prairies at latitude 50°N, while in central Texas at 32°N (and 1,200 miles due south as the crow flies) it may not hibernate at all. The California ground squirrel (*Spermophilus beecheyi*) hibernates from November to January in the milder sea-level regions of Southern California (although young animals may be active all winter), whereas at higher altitudes it hibernates for almost six months.

Although the arrival of winter in the form of overnight frosts and snow certainly encourages the start of hibernation in most species, some do not start immediately. The Olympic marmot (*Marmota olympus*) is one of these, for it may not disappear for several days after the arrival of snow and night frosts, implying that a mechanism other than temperature is the hibernation trigger. As many hibernators awaken and emerge at predictable times it appears that an internal clock (a biological rhythm) controls their sleep patterns, for when they are deep under the snow they cannot possibly know when it is a few degrees milder up above, or that the days are getting longer, certainly not to the extent that they can all appear at almost exactly the same time. The woodchuck, for example, generally arouses 180 days after commencing its hibernation. Their internal clocks must also be the trigger for the white-tailed prairie dogs, which did not appear during warm spells in Wyoming in mid-January one year, but came up as usual at the end of February in the middle of a severe cold spell. There may also be considerable variation according to sex and age in the scheduling of starting and ending hibernation. Adult Richardson's ground squirrels hibernate for seven months, whereas juveniles sleep for just six months, the extra time for adults being due to their earlier commencement in the fall, not their later emergence, and the youngsters have more time to feed and build up their fat reserves. Males also arouse before females, to establish their territories, which is common behavior in the ground squirrels.

Climate change is now also affecting the length of some rodents' sleep. According to a recent study at the Rocky Mountain Biological Laboratory in Crested Butte, Colorado, woodchucks residing nearby are emerging from hibernation earlier than they did thirty years ago, the Rockies having become warmer by 2.5°F (1.4°C) in April since 1976. Conversely, chipmunks and golden-mantled ground squirrels are emerging later, by several days, as their emergence is dependent upon the depth of snow, which has increased by 18 inches (46 cm) in April in the past thirty years. Their evolution during this period has obviously hinged upon an individual's ability to store the extra food required for the longer sleep.

Generally, northern rodents hibernate for six months annually, but several sleep much longer. The Arctic ground squirrel (*Spermophilus undulatus*) is underground for at least eight months, and species that practice estivation that continues into hibernation may slumber for three-quarters of the year. The Uinta ground squirrel is one of these; it begins estivating in July, but may not reappear until the following April, thus spending only about three months above ground. A captive thirteen-lined ground squirrel began estivating on August 4, and did not reappear until March 22, having been asleep for almost eight months. With a combination of cold and fasting, a laboratory dormouse was kept asleep for a year.

Daily torpor also occurs in some of the smaller rodents; like the shrews and hummingbirds, their metabolism drops so they conserve energy. This happens in the pygmy jerboa (*Salpingotulus michaelis*), which lowers its body temperature while sleeping during the day; and the dwarf hamster (*Phodopus sungorus*), in reverse as it is nocturnal. In the laboratory prairie deer mouse (*Peromyscus m. bairdi*) were chilled down to 33.8°F (1°C) for several days and became torpid by day but were active and ate their sunflower seeds at night. The same species that were provided with nest boxes in the wild in Michigan went into a deeper torpor for several days at a time during cold weather, and could not roll over when turned onto their backs. They showed the classic symptoms of hibernation—lowered respiration and heartbeat and body temperature that dropped to 64.4°F (18°C).

It is generally believed that the freezing of the vertebrate's cell contents will result in death, and this occurs when a hibernating animal does not arouse when its temperature drops below the safety line—the set point—probably because its fat stores have been depleted. However, one mammal—the Arctic ground squirrel— does survive the natural lowering of its body temperature below freezing due to the phenomenon of supercooling, when the lack of solid particles within the cells around which ice crystals can form allows the fluids to remain liquid when the squirrel's core temperature drops to 26.6°F (−3°C).

Some of the Species

Marmots

Although squirrels are generally associated with trees, almost half of all the living species are ground-dwellers, and the marmots are the largest of these and are the largest northern rodents. They all belong to one genus, *Marmota*, and are close relatives of the ground squirrels (*Spermophilus*) and the prairie dogs (*Cynomys*), all of which are members of the squirrel family *Sciuridae*. Marmots are totally herbivorous and diurnal and are restricted to the northern areas of both the New and Old Worlds (the Holarctic Zoogeographic Region). There are six species in North America but there is disagreement over the classification of the Eurasian marmots, with some authorities dividing the bobak marmot into four separate species. Marmots are large and stocky animals, which are all deep hibernators that sleep on average seven months each winter, but reaching almost nine months in the Kamchatka marmot (*Marmota camtschatika*) of northeastern Siberia. They do not store food but live off their fat reserves, which may amount to one-quarter of their body weight when they commence hibernation. Their winter burrows are usually about 10 feet (3 m) deep and 65 feet (20 m) long, but much deeper and longer ones have been reported. They are mainly animals of the high country such as alpine meadows and rocky mountainsides (where it is unlikely their burrows can be very deep), plus steppe and forest edges. Their social life varies, from the colonial alpine marmot (*Marmota marmota*) and yellow-bellied marmot (*Marmota flaviventris*) (see color insert), which live in family groups of up to fifteen animals that huddle together in their den for the winter, to the very unsocial and solitary woodchuck (*Marmota monax*). Marmots in central Asia are still a major source of food for the local people and their fat is believed to cure lung disease and rheumatism. They are also shot in large numbers in North America for sport and for their pelts.

Alpine Marmot (*Marmota marmota*)

This is the marmot of the European mountains, including the Alps and Slovakia's Carpathian and Tatra Mountains, where it lives on alpine pastures up to 10,500 feet (3,200 m). It has also been introduced into the Pyrenees where it is now established, and many present-day colonies of this species are the result of restocking areas with animals from Berchtesgaden (later the site of Hitler's aerie), in southern Germany in the nineteenth century. It is a medium-sized marmot, with a brown coat and brownish-orange patches on the back and belly, and a chestnut muzzle; it reaches a body weight of 13 pounds (6 kg).

For about four months in summer alpine marmots graze the vegetation just below the snow line on the warm southern slopes, feeding ravenously and eating far more than they need for daily maintenance. The excess is stored as a thick layer of fat, which can form one-third of their body weight prior to hibernating. They migrate to lower elevations, even into the tree zone, when they have stored sufficient body fat for their hibernation, which lasts from September to early May. However, this varies with the elevation of their burrows, which may be 33 feet (10 m) long and lead into a large den lined with grass, about 10 feet (3 m) below the surface. After plugging the entrance to the burrow with grass, stones, and earth, the family members, often up to fifteen animals, huddle together with their noses between their hind legs. Their metabolic functions are reduced to a minimum, and they are believed to awaken only every four weeks. Their breathing rate drops to two or three per minute with an associated large decrease in oxygen consumption, their body temperature drops from 99°F (37.3°C) to about 41°F (5°C), and their heart rate is only three or four per minute instead of the usual 120 beats. When they emerge they clean out the nest of grass and wastes, and begin their reproductive cycle. Their gestation period is five weeks, and the young appear when about six weeks old in mid-July, having barely two months to grow and store sufficient fat for their first winter; the young sleep with their parents for their first winter. Alpine marmots are totally herbivorous, and eat grasses and herbaceous plants, roots, bulbs, and tubers, and they are fond of flowers. Their major predators are golden eagles and foxes, and they quickly return to their burrows or hide among rocks when they feel threatened.

Alpine Marmot *An alpine marmot rests on the French side of the Pyrenees, where the species was introduced from the Alps and is now established. A body-fat storer, it will hibernate from September to early May, during which time its temperature dips to a few degrees above freezing, and its heart rate and breathing will drop to just two or three times each minute.*
Photo: Courtesy Erin Silversmith

Woodchuck (*Marmota monax*)

The woodchuck or groundhog is the object of an American holiday and certainly the most famous North American hibernator. It has a wide range, from Alaska through most of central and southern Canada, then down the eastern half of the United States to the Atlantic and almost to the Gulf Coast. It lives in open woodland, rocky ravines, brush-covered hillsides, and farmland. It is a heavy-bodied and short-legged animal with a maximum weight of 11 pounds (5 kg), and is usually dark brown with a paler belly, and the whitish tips to the hairs on its back give it a frosted appearance.

Woodchucks are the most solitary marmots, aggressive, intolerant, and so territorial that they chase their young away when they are only six weeks old. Males normally emerge before the females in spring to reestablish and defend their territories, but both sexes retain enough of their winter stores to cope with the needs of territory defense, courtship, and mating, and do not need to find food immediately. After mating, the females return to their dens to continue sleeping, and to conserve energy for the forthcoming birth and raising.

Groundhog Day, on February 2, is an American tradition that originated at Punxsutawney, Pennsylvania, where legend decrees that if the groundhog emerges and sees its shadow it goes back for another six weeks as spring will not be early. If it is a cloudy day, however, and there are no shadows, an early spring is possible. In nature, the woodchuck's actual date of emergence varies. In the north, the hibernation period lasts for at least seven months, but in the Adirondacks they enter hibernation at the time of the autumnal equinox, which is September 23, or the beginning of the fall in the Northern Hemisphere, and emerge to coincide with the vernal or spring equinox on March 21, often before all the snow has melted.

The woodchuck's hibernation nest is at the end of a long burrow, sufficiently deep to deny frost penetration. During hibernation its heart rate drops from the normal 85 beats to about five per minute. Its body temperature drops from 95°F (35°C) to 38°F (3.3°C), and its breathing rate slows to about two breaths per minute, with occasional apnea for several minutes at a time. This lowered metabolism is sufficient to keep the cells functioning. It uses stored body fat for its energy needs, although there have been reports of some woodchucks storing food in their burrow. The woodchuck is aroused by its nervous system if the nest chamber temperature drops, and it wriggles and shivers to warm up. Abandoned woodchuck burrows provide homes for rabbits, weasels, foxes, and opossums.

Bobak Marmot (*Marmota bobak*)

The bobak marmot is an animal of both lowland and high altitude grasslands, ranging from central Europe to the semiarid, grass-covered plains (the steppes) of central Asia, but it is now rare in Europe due to the loss of its habitat to agriculture. It has a golden-brown coat with a darker head and tail, and has a maximum body weight of 13 pounds (6 kg). On the Asian steppes, where it is the equivalent of North America's prairie dog, it is still an important part of the Mongolian economy, for food, medication, and its pelt. There is some risk attached to this close association between humans and marmots as the animal's fleas have transmitted bubonic plague to their hosts, which in turn have infected people. The bobak marmot is a social animal that lives in family dens that usually have numerous tunnels and several entrances surrounded by mounds of soil. These burrows may be 65 feet (20 m) long, with the hibernation chamber 10 feet (3 m) deep to avoid the frost. Their hibernation period is influenced more by elevation than latitude, and in the lowland grasslands their sleep begins in mid-October and lasts until the end of March, but in the high mountain steppes they hibernate from the end of September to early May, having stored so much fat they almost double their normal body weight. Bobak marmots close their burrow entrances with stones and earth to deter wolves, foxes, and bears; but when outside during the summer they are preyed upon by snow leopards, eagles, and ravens. Their gestation period is forty-five days and their normal litter is five, which nurse for thirty days, and appear outside in mid-June at the lower elevations.

Hoary Marmot (*Marmota calligata*)

The largest of the terrestrial squirrels, this marmot grazes herbaceous plants at high altitudes in the Rocky Mountains from Alaska south to Oregon, except along the shores of the Bering Sea where it lives at sea level. Its preferred habitat is talus slopes and rockslides, usually on south-facing mountainsides that give access to alpine meadows. The hoary marmot is silver-gray with a brown rump; it has black-and-white markings on its face, head, and shoulders; black feet, and a large and bushy reddish tail. It has a head and body length of 21 inches (52 cm), its tail is about 8 inches (20 cm) long, and it weighs up to 18 pounds (8 kg). Like all marmots it is a very wary animal, always on the lookout for eagles—and gives a shrill alarm whistle to alert the other members of the group when danger threatens. Bears are also very efficient marmot predators, as they awaken earlier in spring and dig the marmots out of their burrows while they are still torpid, so they often burrow between rocks and then beneath them to deter bears. The marmot has a thick coat and lacks sweat glands; with its layer of hibernation fat, toward the end of the summer it cannot tolerate high temperatures and remains in its burrow on warm days. Marmots are very social animals that live in groups comprising one or more pairs with their young. Their hibernation period varies, from September to April in Alaska and the Yukon; and from October to March in Idaho and Montana. Mating is said to occur within the hibernation burrow, which is rather unusual, as this must be quite difficult and would eventually result in every family group becoming highly inbred, followed by inbreeding depression, poor reproduction, and the likely eventual loss of the population.

Ground Squirrels

There are forty species of ground squirrels, of which most live in North America where they are usually called gophers, and the others in Eurasia where they are known as sousliks. They are all animals of the Holarctic Region—the temperate Northern Hemisphere—ranging south into Mexico and in the Old World to Asia Minor and Iran. Like the marmots, there is disagreement over their classification, and it is expected that the advances being made in genetic studies will soon completely revise the current arrangement. Medium-sized terrestrial squirrels, they are smaller than the marmots and are generally long-bodied and short-tailed. They are animals of the open country, including the North American prairies, tundra,

Eurasian steppes, rocky mountainsides, semiarid regions, and bushy areas adjoining the grasslands. All are diurnal and are expert burrowers that make extensive tunnel systems. The thirteen-lined ground squirrel has a very simple but effective one, as it digs its main burrow, about 12 feet (3.6 m) long, sloping gradually down to its hibernation chamber. Halfway along this tunnel it digs another one at right angles, which deters predators, and which eventually breaks free in the form of a very small clean-sided hole with no tell-tail signs of soil spread around. Badgers and other predators digging into the squirrel's main tunnel entrance are foiled when they reach the right-angled bend because they cannot kick the soil out of an L-shaped tunnel.

Although they are primarily terrestrial, the ground squirrels can climb twiggy bushes to reach nuts and berries. Most hibernate, but there can be considerable variation in the length of their dormancy, even within the same species, depending on the altitude and latitude of their habitat. They store fat for hibernation, almost doubling their weight, but some also store food, which is believed to be used at emergence time when food outside is scarce or still covered with snow. Several species also estivate. The Columbian ground squirrel (*Spermophilus columbianus*) emerges from hibernation in early May, and then estivates for the whole of August, reappearing in September before hibernating again in October. The Mohave ground squirrel similarly hibernates and estivates annually. Ground squirrels vary in their social life, some being quite colonial (Richardson's, Columbian, and Arctic ground squirrels) while others are solitary (Franklin's, thirteen-lined, and California ground squirrels). They are more omnivorous than the marmots, and include insects, small reptiles, and even carrion in their otherwise herbivorous diet; they are opportunistic cannibals that consume their dead companions, and there have been reports of adults killing and eating the young of conspecifics.

Arctic Ground Squirrel (*Spermophilus undulatus*)

The Arctic ground squirrel is a tawny to reddish-brown animal flecked with white, with grayish sides, tawny legs, and a tawny and black tail. It has a head and body length of 12 inches (30 cm), a 5-inch-(13 cm) long tail, and an adult weight of 2½ pounds (1.2 kg). It is the northern-most ground squirrel, which survives on the Arctic tundra where summers are so short there are only about sixty days suitable for plant growth. Its range includes northeastern Siberia, Alaska, and northwestern Canada, within the Arctic Circle and to the shores of the Arctic Ocean to latitude 38°N. This is permafrost country where the soil remains below freezing at all times, except for the shallow surface layer—the active or seasonal frost layer—which supports the scant tundra vegetation, and which thaws each year. The ground squirrels therefore cannot burrow deeply and have evolved a means of combating the subfreezing temperatures from which they cannot escape. They are the only mammals known to practice the phenomenon of supercooling,[2] which allows a liquid to stay fluid at temperatures below freezing due to the lack of solid particles around which ice crystals can form. The animal's cell water therefore remains in a liquid state when its core temperature drops to 26.6°F (−3°C). The Arctic ground squirrel periodically shivers violently for several hours to raise its body temperature back up

Arctic Ground Squirrel   *A rodent of the bleak Arctic tundra, this ground squirrel is one of the few mammals that can withstand the lowering of its body temperature below freezing. While hibernating for eight months it survives this normally fatal situation by employing the physiological phenomenon known as supercooling.*
Photo: Michael Ellis, Shutterstock.com

to the normal 98°F (36.7°C) and then it slowly drops back again. It relies mainly upon stored fat for its hibernation energy, but one burrow that was opened contained seeds, grass spikes, and willow leaves, about a double handful of each, all piled separately close to the squirrel's underground nest. It is not known whether this food is for eating during hibernation or when the animal arouses in spring. The Arctic ground squirrel hibernates from October to May, and the young are born after a gestation period of just twenty-five days. Their growth is very rapid and babies of the year enter hibernation as adults, ready to mate as soon as they emerge in spring.

Thirteen-lined Ground Squirrel
(*Spermophilus tridecemlineatus*)

A long and slender ground squirrel, this species has thirteen alternating lines of tan and dark brown (sometimes broken into spots) on its back and sides, and pale yellowish-brown underparts; it is 11 inches (28 cm) in length, including its 4-inch- (10 cm) long tail. An animal of the grasslands from southern Canada to the Gulf of Mexico, it favors open areas, such as the short-grass prairies, pastureland, and roadsides. Solitary by nature, it lives in a shallow burrow system in well-drained soil, seldom more than 2 feet (61 cm) deep, although it makes a much deeper burrow for hibernation. It has a varied diet that includes seeds, grasses, grasshoppers, and occasionally small reptiles, birds' eggs, and carrion, and in turn is a favored prey of foxes, weasels, coyotes, bull snakes, and red-tailed hawks.

The thirteen-lined ground squirrel really does ensure that it has sufficient food for its long sleep as it is one of the few species that is a combined food and fat storer, laying down a layer of fat in its body to the extent of almost doubling its weight, and carrying food to an underground storeroom. It uses body fat for warming up during its periodic arousals, and then apparently eats stored food, saving sufficient for when it finally arouses in spring, when food is scarce. It is considered a "profound" or "true" hibernator, which has been the subject of numerous research studies into the phenomenon of hibernation. This species enters hibernation within twenty-four hours of being exposed to low temperatures, and becomes torpid very quickly, its body temperature dropping close to freezing, its heart beating only 20 times per minute instead of the regular 200, and its respiration rate drops to 1 per minute from 150 with periods of apnea. It hibernates from October to March or early April in the northern parts of its range, rolling into a tight ball in its nest chamber.

Richardson's Ground Squirrel
(Spermophilus richardsonii)

This is the common ground squirrel of the northern prairies, occurring on the open grasslands from Alberta and Montana eastward to Manitoba and Minnesota. It is often mistaken for the prairie dog, and is very similar, with a buffy-yellow coat, but instead of being black-tipped its tail is brownish with a black tinge. Unlike the solitary Franklin's ground squirrel, it is a colonial animal that lives in complicated burrow systems that may contain several spherical nesting chambers and a number of entrances, the main one being surrounded by a mound of soil. Although this ground squirrel stores food—especially seeds, grain, and dried grass—it is believed this supply is for the awakening and emergence period in spring when fresh food is unavailable, and that it relies upon internal fat stores for energy while dormant. It hibernates for at least six months, from mid-October to mid-April, and males emerge first in the spring to establish their territories, which usually encompass the territories of several females who then establish their own territories after being mated. Suitable habitat may support up to twenty individuals per acre. Richardson's ground squirrel is a major source of food for many predators, including badgers, weasels, snakes, and hawks, especially when the young of the year disperse to establish their own territories, a time when they are very vulnerable.

Gold-mantled Ground Squirrel
(Spermophilus lateralis)

The gold-mantled ground squirrel is an animal of the high country, of the coniferous and mixed mountainside forests, rocky meadows, and sagebrush and

Richardson's Ground Squirrel *A normal, buffy-colored Richardson's ground squirrel and its mate, an albino mutant, on the northern prairies, where the severe winters force them to hibernate from October to late March. They rely mainly on stored body fat for energy during their torpidity, but may also have a small store of dried grass.*
Photo: Clive Roots

chaparral of the Rocky Mountains, the Cascades and the Sierra Nevada, from southern Canada to Arizona. It is one of the smallest ground squirrels, just 8 inches (20 cm) long, and is often mistaken for a large chipmunk, but differs from that group of squirrels in the absence of facial stripes on its coppery-colored head. Its back is striped, however, with white bordered with black. The gold-mantled ground squirrel is a burrower, and its hibernation period varies with the altitude of its habitat. In the Canadian Rockies it starts its long sleep at the end of October and then emerges in May, whereas in the Sierra Nevada it hibernates from early November to the end of March. Its summer burrows are short, about 18 inches (46 cm) long and 10 inches (25 cm) deep, usually with the entrances carefully hidden behind a rock or a log. For hibernating, it digs a much longer and deeper burrow, which has short blind passages that are used as food caches or toilets. It stores fat internally and gathers nuts, fungi, and berries for its cache; it is consequently considered an atypical ground squirrel hibernator, as its frequent periods of arousal to eat are a chipmunk characteristic (see color insert).

Prairie Dogs

Prairie dogs are the most social squirrels and are close relatives of the marmots and ground squirrels. They are stocky, short-legged animals that have a very important role in the ecology of the grasslands, turning the soil, improving drainage, providing homes for many other species in their burrows, and supporting a wide range of predators. There are five species of prairie dogs. The black-tailed prairie dog (*Cynomys ludovicianus*) lives on the dry plains from southern Canada to central Texas. The white-tailed prairie dog (*C. leucurus*) occurs at high altitudes in Colorado, Montana, Wyoming, and Utah. Gunnison's prairie dog (*C. gunnisoni*) has a limited range in the Four Corners region of the United States, from 5,000 feet to 11,000 feet (1,525 m to 3,350 m). The Mexican prairie dog (*C. mexicanus*) is restricted to northern Mexico, where it is now very rare; and the Utah prairie dog (*C. parvidens*), which is the smallest species, occurs only on the short-grass prairies of south-central Utah.

Their periods of hibernation vary considerably. The Mexican prairie dog is active all year except at the higher elevations of its range, and even breeds in January, but the other species sleep for varying periods during the winter, depending on the latitude or elevation of their habitat. However, it has become fashionable to say that the black-tailed prairie dog does not truly hibernate, but just has shallow and infrequent periods of torpor, as it awakens from its sleep and appears regularly during the winter months. Although this is true of the southern parts of its range, it certainly is not on the northern prairies. During the 4–5 months of severe winter in central-southern Canada, where temperatures dip to −31°F (−35°C) regularly and for days at a time (and occasionally to −40°F (−40°C) in December and January), prairie dogs unquestionably hibernate. The members of a captive colony in this environment were induced to sleep by the arrival of low temperatures, and only very occasionally did one appear on milder days in midwinter, when the temperature was just below freezing. The lowest temperature for an appearance was 21°F (−6°C), so they were out of sight in their tunnels for several months. The tremendous increase in body weight due to fat deposition prior to hibernation leaves no doubt as to the source of energy during their sleep. It has also been suggested that the black-tailed prairie dog may have a normal body temperature during hibernation, but this was disproved by a team from Colorado State University, which monitored the temperatures of five wild individuals during the winter and spring of 1998/1999. All became torpid, their bouts of torpidity varying from major ones lasting an average of six days, when their body temperature dropped to 66.2°F (19°C), to minor ones of about thirty hours when their temperature rose to about 89.6°F (32°C), still well below normal.

Black-tailed Prairie Dog (Cynomys ludovicianus)

This is the most familiar of the terrestrial squirrels, which once existed in the millions on the short-grass prairies, living in "dog towns" that covered up to 1,000 acres (405 ha). One huge colony in western Texas early last century was said to contain several million animals. Their enormous burrow systems are subdivided into smaller units ending up at the family level, but there is limited social connection between the groups, and they defend their territories. Development of the prairies for grazing and the additional short-grass habitat this produced is believed to have initially created a huge increase in prairie dog numbers and they were then poisoned as their competition with cattle for the grazing rights became unacceptable. This has reduced them to less than 1 percent of their earlier numbers, although they are still locally common animals.

The black-tailed prairie dog is about 16 inches (40 cm) long, weighs 3 pounds (1.5 kg), and is pale cinnamon-buff with a buffy-white belly and black-tipped tail. It has large eyes and very good vision. Although it carries grass down to its hibernation chamber, this is for nest-making as the prairie dog does not eat during its sleep, but relies on stored fat which is so extensive it almost doubles the animal's normal weight. The prairie dog towns are important ecologically, providing food for many animals, including foxes, coyotes, diurnal raptors and badgers, and black-footed ferrets before their virtual demise. Rattlesnakes and burrowing owls share their dens, which also provide homes and hibernation sites for several species of reptiles and amphibians. In cold zones such as the prairies of North Dakota and Manitoba, prairie dogs spend the winter in their burrows, deep beneath the snow and the frost line from mid-November to mid-March, and reappear only rarely on mild days; this frequency increases farther south as the winters become shorter and milder days more frequent. Mating occurs soon after their arousal from hibernation and after a gestation period of thirty-five days, the young begin to appear above ground in mid-June when they are five weeks old. This type of sleep is called facultative torpor, in which the starting time is optional and depends upon the conditions, being prompted by shortage of food.

White-tailed Prairie Dog (*Cynomys leucurus*)

This is a less colonial species than the black-tailed prairie dog, and within its colonies only a few of the burrow systems are linked. It is also a stocky animal, its basic coloration being pinkish-buff mixed with black above, and paler below, with a short, white-tipped tail. It lives in the high mountain valleys, open country, and brush land from 5,000 feet (1,525 m) to 8,500 feet (2,600 m) in the Rocky Mountains from Montana south to New Mexico. The white-tailed prairie dog is known as a spontaneous hibernator, which enters torpor in October regardless of the weather or food availability, and emergence in spring is also independent of weather conditions. Normally it appears at the end of February or in early March, but in January 1965, despite a mild spell when temperatures rose to 60°F (15.5°C) and all the snow had melted, no prairie dogs appeared. They eventually emerged as usual at the end of February, having to tunnel through 6 inches (15 cm) of snow to reach the surface, where a severe cold spell had sent the temperature below freezing. As in most of the terrestrial squirrels, males normally enter hibernation before the females and emerge sooner. On the Laramie Plains in southeast Wyoming, males have emerged in early March, three weeks before the females. Mating occurs as soon as the females emerge and the babies appear in mid-June, at the age of six weeks. This species has also been known to estivate in July.

Chipmunks

Chipmunks are mainly animals of the forested and rocky regions of North America and northern Eurasia, especially at higher elevations, in both boreal and mixed forests, on brush-covered mountainsides and in chaparral, but they do not occur above the tree line, and are quite strict at staying within their altitudinal ranges. In the Sierra Nevadas, the lower mountain slopes and the chaparral are home to Merriam's chipmunk (*Eutamias merriami*), up to about 6,000 feet (1,830 m). The lodgepole chipmunk (*E. speciosis*) lives on the western slopes from 5,000 feet (1,525 m) to 11,000 feet (3,350 m), and the Uinta chipmunk (*E. umbrinus*) occurs on the eastern side at the same elevation. The two are separated by the alpine chipmunk (*E. alpinus*), which lives on both slopes of the peaks up to 13,000 feet (3,960 m).

Chipmunks are active and alert animals with high metabolic rates, and are much smaller than the ground squirrels, ranging in size from the tiny least chipmunk (*Eutamias minimus*), which weighs 1½ ounces (42 g), to the eastern chipmunk (*Tamias striatus*) at 4 ounces (112 g). They are distinguished by the stripes on their sides and faces and their habit of running with tail held vertically. There is disagreement over their classification, but of the twenty-five or so species all are North American except one—the Siberian chipmunk (*Eutamias sibiricus*)—and all species hibernate for varying periods depending on the elevation or latitude of their home range.

Chipmunks eat a wide range of foods, and gather and store large quantities for their hibernation, stuffing it into their cheek pouches, and then pushing it out into their underground storage chamber by the action of their paws against the cheeks. They favor seeds, especially those of jack pines and lodgepole pines, and almost 800 jack pine seeds were once counted among the contents of a chipmunk's stomach. When pine seeds are scarce they turn to fungus, especially the underground species *Gasteromycetes* and *Tuberales*, which they locate by their strong odor. As these are perishable, they eat them where they find them rather than carry them to their stores. They also eat and store acorns, the seeds of manzanita, deer brush, and the mountain whitethorn, and berries of the choke cherry, blueberry, and pin cherry. They are careful to store only dry foods. They eat insects, especially caterpillars, and are a major control of the larvae of an *Ethmiid* moth, which is a pest of the desert mahogany tree, and they also eat birds' eggs and nestlings. Chipmunks dig their own burrows, the entrances to which are usually concealed beneath rocks, tree trunks, or bushes, and they may be 13 feet (4 m) long and have a number of branches, with a chamber at the end of each, for sleeping or food storage.

Hibernation for most chipmunks begins in October or November and lasts until March or April, but dormancy depends upon their habitat, and the gray-necked chipmunk (*Eutamias cinereicollis*) of Arizona and New Mexico may not hibernate at all. Their metabolism drops drastically while they sleep, with their heart rate falling to about 15 beats per minute and their temperature dropping from 102°F (39°C) to 37.4°F (3°C), which results in a 75 percent saving of energy.

As they do not store fat to fuel even this reduced energy requirement they must awaken periodically to eat, defecate, and urinate, but still lose about half their body weight during the winter.

Chipmunks appear on the surface on warm days, and also store food outside their burrows, which they search for when they emerge in spring. Males appear first and females a week later, when breeding begins almost immediately. Some species may have two litters per year, in spring and late summer. Their gestation period is about thirty days and the young are born naked and blind and appear at the entrance to their burrows when they are about one month old. They are weaned at six weeks and are able to breed the following spring.

Siberian Chimunk (*Eutamias sibiricus*)

This chipmunk inhabits the taiga—the moist coniferous forest that borders the tundra—and the adjacent mixed coniferous and deciduous forest of northern Eurasia from Scandinavia to far-eastern Siberia. It prefers rugged, rocky forest-clad areas and brush-covered mountainsides in this region, which experiences extremely cold winters and very hot summers. The Siberian chipmunk is slightly smaller than the North American eastern chipmunk, being just 5½ inches (14 cm) in head and body length, and has a tail about 3 inches (7 cm) long. It has pale buffy-white and brown facial and body stripes on an ochre-yellow or grayish-brown base color—depending on its locality and the subspecies—of which there are several in its wide range. It has four light and five dark longitudinal body stripes and its tail is light brown with black lines on the sides. Unlike the many American chipmunks of the same genus, the Siberian chipmunk stores its winter supply of food in a separate chamber, not in its nest. It is a good climber, although primarily a ground-dweller, and burrows to nest and to hibernate, which occurs from October to late March. It is a typical omnivore, eating nuts, berries, buds, fungi, birds' eggs, and nestlings, and when it lives near habitations it raids cereal crops of wheat, buckwheat, and oats.

Eastern Chipmunk (*Tamias striatus*)

There are numerous chipmunks in North America but they all belong to the genus *Eutamias*, with the exception of this species, the only member of the genus *Tamias*. It is slightly larger than the others, about 6 inches (15 cm) in head and body length with a 4-inch-(10 cm)-long tail, and weighs almost 4 ounces (113 g). Reddish-brown in general color, it has five dark stripes alternating with grayish-brown ones, and buffy-white underparts. It occurs only in the eastern half of North America, from central Canada south almost to the Gulf of Mexico, but excluding Florida. The eastern chipmunk is an animal of the deciduous forests and rocky scrub regions, and with such a wide latitudinal range its hibernation period varies considerably. In northern Quebec it sleeps from October to early April, but may appear on warm days, whereas in the extreme south of its range it has remained active all winter. In addition, there is considerable individual variation in the

Siberian Chipmunk   *Eurasia's version of the familiar American rodent, this species lives in the vast northern forests that stretch from Scandinavia to Russia's Pacific coast. Its long winter sleep (of five or six months duration) is fueled by foods cached in an underground storeroom, and it awakens every few days to eat.*
Photo: WizData.inc., Shutterstock.com

periods of dormancy in this species, and its frequency of awakening to eat is unknown. A solitary animal, it lives alone in its own underground burrow, and stores food for the winter, but unlike the other American chipmunks that store food in their nest chambers, it has a more extensive burrow system, which includes a sleeping compartment and separate storage chambers. It fills these through making many trips with food held in the cheek pouches that open into the sides of its mouth. It has a fairly typical omnivorous diet, composed of seeds, nuts, berries, fungi, and bulbs, plus insects, birds' eggs, and nestlings. Although terrestrial it climbs trees readily in search of food.

Least Chipmunk (*Eutamias minimas*)

This is the world's smallest chipmunk, and also the species with the widest altitudinal and geographic range. It is less than half the size of the largest species, with a head and body length of 3½ inches (9 cm) and a bushy tail of the same length; it weighs just 1½ ounces (42 g). It has three dark and three light stripes on its face, and five dark and four light stripes on its sides, the middle stripe extending to the end of the tail. Its background color is orange-brown on the back and grayish-white on the underparts. The least chipmunk is mainly a forest animal, occupying northern forests dominated by coniferous trees, and mixed deciduous forests, mostly at high altitudes, although it prefers more open places such as clearings and the forest edge. It also occurs in sagebrush flats, rocky cliffs, and river bluffs, and even the badlands of South Dakota, where it sleeps at night, and hibernates, in crevices in the sandstone outcroppings. It has a wide range from New Mexico north through the Rocky Mountains to Canada's Yukon Territory, extending eastward to Wisconsin and the Great Lakes and north from there to Hudson's Bay. Hibernation takes place in an underground burrow which it digs beneath logs, rocks, or tree stumps, spending several months there in a nest of grass, feathers, and shredded bark, and awakening periodically to raid its larder. In most places within its range, influenced by high latitude or high elevation, it escapes the harsh winters by hibernating from October to April. Although terrestrial, the least chipmunk is a very good climber and may even make a summer nest in a tree. It searches for buds, nuts, berries, flowers, fungi, and insects such as beetles, grasshoppers, and caterpillars; in turn it is preyed upon by snakes, red-tailed hawks, weasels, mink, and feral cats.

Flying Squirrels

There are about thirty-six species of flying squirrels, which vary in size from the pygmy flying squirrels (*Petaurillus*), which weigh about ⅔ ounce (20 g), to the giant flying squirrels (*Petaurista*), which weigh over 4 pounds (2 kg) and are the largest of all the tree squirrels. The majority of these rodents are warm-climate animals and do not need to hibernate. Only four species living in the northern temperate regions, two Old World flying squirrels (*Pteromys*) and two New World flying squirrels (*Glaucomys*), are known to become torpid in winter. These small squirrels are rarely seen as they are strictly nocturnal and totally arboreal, and hide in natural tree crevices and hollows during daylight. Despite their name, however, they have no flapping capability at all and do not fly like the bats and birds; their "wings" are a membrane called the patagium which extends between the limbs along the sides of the body. These open out like a blanket to increase the surface area and allow the squirrel to glide down from a high point.[3] Direction control is provided by membrane-adjusting muscles and by angling their well-furred tail like a rudder. A normal glide is about 10 feet (3 m), but one of 90 feet (27 m) has been observed, and just before reaching their landing place they turn upward, to alight on the tree trunk on all fours. The flying squirrels make high-pitched calls while gliding, which may have some echolocation value to guide their approach and touchdown.

Northern Flying Squirrel (*Glaucomys volans*)

One of the two native North American flying squirrels, this species occurs across Canada south of the tundra zone and in the northern United States, where it lives in the mixed deciduous and coniferous forests. It is almost double the weight of the other native species, the southern flying squirrel (*G. sabrinus*), and the hairs on its belly are white only at their tips rather than for their whole length. Normally solitary, *Glaucomys volans* is a communal hibernator in which as many as twenty animals may congregate in the same winter nest, usually a tree cavity, presumably to benefit from the warmth they generate. But they are not deep hibernators, and in fact are quite cold-hardy, so they do not become torpid and remain in their nest until the temperature drops to about $-10°F$ ($-23.3°C$). Their main winter diet is believed to be arboreal lichens, but they also feed from food stores, possibly their own, although they are known to steal from red squirrels' caches. It seems likely however, that they feed from these stores during their normal periods of activity, and not during arousal from torpor during extreme cold, so they must therefore have sufficient reserves to provide energy for these dormant periods. Their main predators are owls, martens, and raccoons. North American flying squirrels are sexually mature at eighteen months, and breed twice yearly, their litters containing up to five young, which are weaned at fifty-six days. Their gestation periods differ, however, being about forty days for the southern species whose young are born in June and July, but only thirty-two days for the northern squirrels, which give birth in May and June, therefore providing more time for them to prepare for winter.

Old World Flying Squirrel (*Pteromys volans*)

A squirrel-like animal, this species is smaller but very similar in appearance and habits to its New World relative *Glaucomys*, with a flying membrane or patagium between its forelimbs and hindlimbs on each side of its body, the margins of which have a fringe of soft fur. *Pteromys* has large eyes and small ears, and a small, flat and furred tail. In flight it extends the forelimbs laterally, keeping the hindlimbs together close to the tail, which produces a triangular silhouette from below. Its coat is long, thick, and very soft and in winter is silvery-gray, which changes to blackish-gray in summer, although their bellies remain white all year. Its habitat is seasonally cold mixed forest of spruce, birch, and conifers; it prefers old growth woodland with mature trees and tree cavities suitable for nesting and hibernating, and which provide its food that includes nuts, pine seeds, pine needles, and catkins. This species ranges across northern Eurasia from Finland to Korea and Japan, and south along the Pacific coast of northern China. It is also strictly nocturnal and arboreal, and usually jumps between branches and glides only across larger spaces. It rarely comes down to the ground, where it is very clumsy, and hides during the day in a tree hole or in a nest which it builds at the junction of a side branch and the main trunk of a fir tree. In winter it becomes torpid during extremely cold weather and remains in its den.

Hamsters

There are about twenty species of hamsters, but only the golden hamster is a familiar animal, the most popular rodent pet despite its very recent domestication.

Hamsters are all animals of the northern temperate regions, including central and eastern Europe, Asia Minor, and central Asia east to the deserts of China. They are the world's great food storers, and with one exception (the mouse-like hamster *Calomyscus bailwardi*) they all have cheek pouches, which are associated with the collection of food and its transportation back to a safe storage place. Storing food externally, like the chipmunks, is in turn associated with hibernation, as it is one of the two methods through which animals provide the energy they need to survive their periods of dormancy (the other being internal fat storage). In the hamsters these vary in length from the daily torpidity (plus longer periods of inactivity during very cold weather) of the nocturnal dwarf hamsters (*Phodopus*) to longer periods of dormancy in the rat-like hamsters (*Cricetulus*), and then the lengthy hibernation of the common hamster (*Cricetus cricetus*), which sleeps from late October to early April in Siberia. Except when in daily torpor, all hamsters awaken frequently to replenish their energy from their food stores. The frequency and length of these arousals in the wild is unknown, but golden hamsters kept at low temperatures in the laboratory remained torpid without awakening for twenty-eight days.

Common or Black-bellied Hamster (*Cricetus cricetus*)

Mainly nocturnal or at least crepuscular, this species is the largest hamster, a guinea pig–sized animal about 12 inches (30 cm) long and weighing up to 2 pounds (900 g). It has a thick coat of light-brown hair with a black belly and white patches on its cheeks and sides, and a tiny naked tail. Albinos and melanistic individuals are often seen. The common hamster occurs from central Europe to central Asia, although it is now very rare in the western parts of its range. It has large cheek pouches, which it inflates with air when it swims (which it only does under duress). It is a solitary and grumpy animal; a male only enters a female's territory for mating and is driven back out immediately afterward. The common hamster lives on the open steppes and along riverbanks, in burrows of varying depths, but reaching 6 feet 6 inches (2 m) deep to escape the frost when hibernating. Like the other hamsters, it stores food for the winter, and needs a large supply to maintain it for several months. Storage chambers holding 200 pounds (90 kg) of seeds, grain, pulses, corn, potatoes, and fungi have been excavated, and in fact purposely raided by humans, to steal the stores. With such gathering capability it is not surprising that this hamster is considered a pest of cereal and root crops. In addition to storage larders, the burrow has a sleeping chamber and a separate compartment used as a latrine. It begins its sleep in October in central Asia, emerges in March, and breeds in early April, but its hibernation is intermittent, for it arouses to feed, urinate, and defecate. It is a very prolific animal, producing up to twelve young per litter after a gestation period of only nineteen days. The young are weaned when three weeks old and are sexually mature at the age of two months. Its numbers are controlled by many predators including weasels, stoats, foxes, badgers, and owls.

Golden Hamster (*Mesocricetus auratus*)

There are actually four species of animals called golden hamsters, which occur across Asia Minor and eastern Europe, but only the popular pet animal *Mesocricetus auratus* is well known. It has a very restricted distribution in the wild, living on the dry, rocky slopes of northwestern Syria, where a female and her four babies were caught in 1930. These were first bred as laboratory animals before being "discovered" by the pet trade. The wild golden hamster has a reddish-gold back and grayish-white belly, but many color and pelage mutations have been produced during its recent domestication. It is a solitary and antisocial animal which is very territorially minded and objects viciously to intruders; some domesticated pets have reverted to this behavior, becoming wild and unreliable when neglected. The female is larger than the male, weighs about 5 ounces (140 g), and is so aggressive that the male must depart quickly when mating is completed. This hamster is a very loose-skinned animal, with huge cheek pouches that extend back over its shoulders, which are used for collecting and carrying food back to the hibernation den within the extensive burrow system. Mother hamsters have been known to hide their newborn young in their cheek pouches when they feared for their safety. The golden hamster is a very productive species, a valuable characteristic in the laboratory and for pet breeders. Litters of up to twelve young are born after a gestation period of only fifteen days and the babies mature rapidly and are weaned when nineteen days old, allowing the female to be mated soon afterward. Hamsters can breed when they are just one month old, so their reproductive cycle is therefore just over six weeks, which is very short for a placental mammal.

Dormice

Dormice are small, squirrel-like animals with soft coats and mostly bushy tails, short legs, and curved claws for climbing, and they are all nocturnal and mainly arboreal. They are members of the family *Gliridae*, which contains nineteen species, but ten of these belong to the genus *Graphiurus* which live in sub-Saharan Africa. Little is known of their natural history although it is believed that the South African members of the genus hibernate for part of the winter. The other dormice are all animals of the Palaearctic Region (the northern Eurasian temperate zone), which hibernate and are mostly internal food storers. They gain weight in the fall with fat stored for their hibernation, which in the north lasts from October to April. The forest dormouse (*Dryomys nitedula*) and the fat dormouse (*Glis glis*) also store food externally.

Dormice are animals of the deciduous forest zone, where they can find beech mast, chestnuts, hazel nuts, and acorns, all very important foods for their regular diet and for producing the thick layer of fat that provides their energy when asleep, whereas coniferous forests lack such variety. They are all nocturnal and prefer dense low undergrowth where the foliage shields them from owl attack at night, and although primarily arboreal, they generally hibernate underground in burrows of their own digging or in the abandoned holes of other animals. Shortening day length triggers the deposition of fat, and they are profound hibernators, with lowered metabolism that chills their bodies down almost to freezing and considerably lowered respiration and heartbeat. Internal body temperature and respiration are connected in hibernating dormice, as they are also in all other mammalian hibernators. When their temperature is below 53.6°F (12°C), breathing may cease (apnea) for several minutes at a time, and as their temperature rises, the periods of apnea lessen until at 68°F (20°C), they are breathing normally and begin to awaken.

Hazel or Common Dormouse (*Muscardinus avellanarius*)

The hazel dormouse is indigenous in the United Kingdom and Eurasia, ranging from France eastward across central Europe to Russia and south to Asia Minor. A woodland animal, it prefers the thicker vegetation of secondary-growth forests, hedgerows, and thickets. It is very agile and resembles a large, but thick-set house mouse, yellowish-brown above with a creamy-white chest and buff belly. It is only 3½ inches (9 cm) long, with a short-haired tail about 2¾ inches (7 cm) in length. The hazel dormouse is exclusively nocturnal, but is less solitary than the fat dormouse and sometimes lives in small colonies. It builds a globular nest in a tree cavity or makes a nest of grass, leaves, moss, and bark in the open branches of a tree, often using an old bird's nest as a foundation. For hibernating, however, it usually makes a nest of moss between tree roots or in a pile of leaves or stones. Its nest material is "glued" together by an excretion of its salivary glands. The hazel dormouse hibernates from late October to April in England, and the trigger for hibernation appears to be solely temperature, for it disappears when it drops below 59°F (15°C). It experiences a very rapid drop in its own temperature when it begins its sleep, from 95°F (35°C) down to just above that of the environment. As its name suggests, this dormouse is very fond of hazelnuts, which it cannot crack, so it must chew a hole in the side to reach the kernel. It also eats beech mast, acorns, buds, berries, insects, and birds' eggs. It stores food as well as fat for the winter.

Fat or Edible Dormouse (*Glis glis*)

This small, bushy-tailed rodent lives in Europe and across central Asia to the Volga River and Iran, and has been introduced into England where it is now established in the south. It prefers deciduous woodland, mainly with low trees and bushes, but also lives in gardens and orchards, and often invades the lofts of country buildings. The dormouse's natural diet comprises nuts, berries, fruit, acorns, birds' eggs, and nestlings; and it causes great damage in orchards and vineyards. It has short, thick, silvery-gray to brownish-gray fur, with a white belly and a dense bushy tail, and has prominent eyes and long vibrissae. In keeping with its nocturnal habits it has a good sense of touch, and its night vision and hearing are well developed. It is the largest of the dormice, with adults measuring 8 inches (20 cm) excluding their long tail, and weighing up to 6.3 ounces (180 g), but they more than double this

Hazel or Common Dormouse *Resembling a stocky mouse with a furred tail the hazel dormouse is a woodland animal of Europe and Asia. It hibernates from late October to April in England, and its long sleep is fueled by body fat, plus a store of its favorite foods—hazel nuts, beech mast, and acorns.*
Photo: Josef Hlasek

when ready for winter. The Ancient Romans fattened them with chestnuts and acorns, either in enclosures called gliraria or in small earthenware pots, and when ready for the table they were so fat they could barely move. They were a food of the rich, and were considered such a luxury that at one time their consumption by the general populace was banned. Their ability to store so much fat stems from their natural hibernation physiology, providing food for their long dormancy. When dormant from October to May they curl into a ball, often in the company of several others, in a nest of grass in cellars, holes in walls and trees, and in birds' nest boxes, but most often underground, burrowing down about 3 feet (90 cm). They awaken for short periods every twenty-four hours. Fat dormice are quarrelsome little animals; in the wild they mate and then go their own way. They are seasonal breeders, having just a single litter annually containing an average of five young, born after a gestation period of thirty days, and weaned when one month old. They are extremely long-lived and captive dormice have reached the age of fourteen years.

Kangaroo Rats and Their Relatives

These small and unusual rodents are members of the family *Heteromyidae*, which contains about sixty species and is restricted to the New World, from Canada south through the United States to Central America and northwestern South America. They are often collectively called heteromyids, and members of three of the six genera are hibernators which occur in the more arid regions of western North America. Of these, the pocket mice (*Perognathus*) resemble mice and act like them too, as they run and do not hop on their hindlimbs. The others—the kangaroo mice (*Microdipodops*) and the kangaroo rats (*Dipodomys*)—have short forelimbs and large kangaroo-like hindlimbs modified for jumping.

The heteromyids have evolved several adaptations for desert survival. They are nocturnal and hide in their burrows during the heat of the day. They do not drink, and obtain some of their fluids from succulent plants, but they also produce their own water through body metabolism, so they can live on a diet of seeds. They select seeds for storage based on their water content, and store dry ones in humid burrows to increase their moisture. During very hot weather they attempt to eat more insects, which have a high preformed water content. They have also evolved renal mechanisms that actually conserve water. Their kidneys are four times more efficient than human kidneys and therefore use little water to remove their nitrogenous wastes, producing small amounts of highly concentrated urine and very dry feces. They have large, fur-lined external cheek pouches that open on either side of the mouth and extend back over the shoulders, for collecting food in good times, which they store for the bad times.

The pocket mice, named for their fur-lined cheek pockets or pouches in which seeds are carried back to the winter stores, are the smallest heteromyids, with several species[4] measuring only 2½ inches (6 cm) in head and body length and weighing just ¼ ounce (7 g); all qualify as the second-smallest terrestrial hibernators (with the little pygmy possum), weighing just 1 g more than the smallest species, the birch mouse. The kangaroo mice are slightly larger, at 3 inches (8 cm) long and weighing ⅓ ounce (9.5 g); and then come the kangaroo rats, which range from 4 inches (10 cm) to 6½ inches (7 cm) in length. The heteromyids are inactive in cold weather. Northern species such as the little pocket mouse (*Perognathus longimembris*), the Great Basin pocket mouse (*Perognathus parvus*), and the dark kangaroo mouse (*Microdipodops megacephalus*) hibernate from November to March; but the silky pocket mouse (*Perognathus flavus*), which ranges from Nebraska to central Mexico, appears to undergo only short-term torpidity for a few days at a time, and has been active above ground even at temperatures down to 14°F (−10°C).

Little Pocket Mouse (*Perognathus longimembris*)

This, tiny soft-furred and large-headed mouse is buffy-yellow above interspersed with black hairs, and its belly is brownish-white. It is 2½ inches (6 cm) long with a 3-inch-(8 cm) long tail and weighs ¼ ounce (7 g); it is one of several species of pocket mice that are second only to the birch mouse as the smallest terrestrial mammalian hibernators. It has difficulty thermoregulating; it gains or loses heat rapidly due to its small size and high metabolic rate, and it enters a state of torpidity with one rapid decline of body temperature, which can be achieved within four hours. The temperature of hibernating mice is close to that of the environment and fluctuates between 39.2°F (4°C) and 68°F (20°C), whereas their normal temperature varies between 89.6°F (32°C) and 102.2°F (39°C). It is a very abundant little rodent, which occurs from Oregon and Utah to Mexico, where food storage, burrowing, and the ability to become torpid when necessary have enabled it to survive the inhospitable extremes of the North American deserts. It hibernates underground and awakens periodically like the chipmunks to feed from its stores.

In California the little pocket mouse is inactive from November to January, and in Idaho it is not seen from October to March. However, it often has periods of dormancy that have no strict seasonal pattern; its potential for daily torpidity exists throughout the year, and may be triggered by food shortages rather than cold weather as hibernation and estivation have been induced in captive mice by starving them. During daily torpor for about ten hours its body temperature drops to 59°F (15°C).

Dark Kangaroo Mouse
(*Microdipodops megacephalus*)

There are two species of kangaroo mice—one dark and one pale—and both are tiny animals midway in size between the pocket mice and kangaroo rats. The dark kangaroo mouse occurs in Oregon, California, Nevada, and Utah, in sandy and gravelly regions, sagebrush, and alkali sinks. It is blackish-gray above and its underparts are grayish-white; it is 7 inches (18 cm) long and half of that is tail, which is unusual in being thicker in the middle than at either end. Although it can walk on all fours, its main method of locomotion is bipedal hopping, like a kangaroo. It becomes torpid when food shortage is combined with temperatures below 50°F (10°C), as a means of conserving energy while maintaining its body weight. It is active from March to early November, during which time it collects a store of seeds, and then spends the rest of the year underground in its burrow. Its degree of torpidity and frequency of awakening are unknown, but in the laboratory it has regulated its periods of torpor to conserve its store of seeds when it was very cold, and these periods varied from a few hours to several days. The dark kangaroo mouse digs unbranched tunnels up to 4 feet (1.2 m) long, although they are rarely more than 1 foot (30 cm) deep. Its diet consists primarily of small seeds and insects, and for fluids it relies totally upon the metabolic water produced from its food.

Ord's Kangaroo Rat (*Dipodomys ordii*)

With their well-developed hindlimbs and bounding gait the rodents of the genus *Dipodomys* resemble miniature kangaroos. Most species are nocturnal burrowers and spend summer days in the microclimate of their burrows to escape the intense heat, plugging the entrance to help maintain an acceptable temperature and humidity. Some are also forced to remain in their burrows for long periods when temperatures are below freezing or when there is a heavy snowfall. Ord's kangaroo rat is one of these, an animal of the sandy regions and areas of dry and hard-packed soil of western North America, from southern Canada down through the western United States to Mexico. It is about 10 inches (25 cm) in total length with a tail 6 inches (15 cm) long, and it has a buffy-red coat and a white belly, with conspicuous white spots above and below its eyes and under the ears. Its tail is dark on the top and bottom and white on the sides, and is tufted along the end third of its length. In Canada and Idaho it hibernates from November to March, whereas in Texas some individuals are active all year. It makes a shallow burrow with long tunnels connecting its nest area with the food-storage chamber, and awakens periodically to feed. Seeds form the bulk of its diet, especially those of mesquite and tumbleweed, plus buds, shoots, and fungi. Like the other kangaroo rats it eats perishable food items when out foraging, and pushes seeds into its large cheek pouches to eat later or to store. While outside its burrow it is preyed upon by snakes, owls, weasels, foxes, and coyotes, and to escape it bounds bipedally on its hind toes with the heel of its foot off the ground, whereas when it walks slowly all four feet fully touch the ground.

Jumping Mice and Birch Mice

The family *Zapodidae* contains twelve species of tiny rodents known as birch mice, which have a wide distribution across eastern Europe and central Asia, and five species of jumping mice, four of which occur in North America. These mice are close relatives of the jerboas and have on occasion been included in the same family, *Dipodidae*, but there are differences in the anatomy of their hind feet, and most authorities separate them. They all have long hind feet and long tails, and have varying degrees of saltatorial locomotion—moving mainly by jumping. Together with their long feet the hindlimbs are modified for bounding in the jumping mice, but less modification has evolved in the birch mice, and they normally move on all fours, almost in a crawl, and by making short hops, although they can leap high and long when alarmed. All the known species of birch and jumping mice are nocturnal, and hide by day under logs or in vegetation, but they also dig their own burrows. They all hibernate for the winter months, relying on their fat reserves for their energy. None are believed to store food externally.

Birch mice are 2 inches (5 cm) in head and body length and weigh only ⅕ ounce (5.6 g). They have long and semi-prehensile tails which they wrap around branches when they are climbing, and use for balance when leaping between branches. They occur in a wide range of habitat including both forest and steppe, and have an omnivorous diet, eating seeds, berries, and insects. They are nocturnal and hide during the day in shallow burrows of their own digging.

The three species of North American jumping mice of the genus *Zapus*[5] are profound hibernators, which sleep from early October to late April in the north of their range, in Alaska and northern Canada. They lack cheek pouches and do not store food in caches, but become very fat in the fall. During their hibernation underground their temperate drops to 35.6°F (2°C) from the normal 98.6°F (37°C). The fourth native species—the woodland jumping mouse (*Napaeozapus insignis*)—lives in eastern North America from Manitoba to the Atlantic Ocean and south into Georgia. It hibernates from October to April in Manitoba and Wisconsin, and also relies totally on its store of fat for energy while asleep.

Southern Birch Mouse (*Sicista subtilis*)

The birch mice are indigenous only to the Old World, where they occur throughout eastern Europe and then across central Asia to the region of Lake Baikal. The southern birch mouse is one of twelve species within the genus *Sicista*, which are considered close relatives of the North American jumping mice. It is dark brown above with a black dorsal stripe along the spine, and has pale-brown underparts. A good climber and jumper, it uses its long and semi-prehensile tail when it climbs in bushes, and holds it out straight for balance when making long leaps between branches. It is active only at night, and is an animal of varied habitat, being found in forests, meadows, moorland, and steppe. It is the smallest terrestrial[6] mammal to hibernate, a tiny creature with a head and body length of just 2 inches (5 cm) and a weight of ⅕ ounce (5.6 g), which doubles in late summer as it accumulates fat for its long winter sleep (it does not store food externally). Winter is passed in a nest of grass in a shallow burrow, and in the most northerly parts of its range its period of torpor may last from mid-September to April.

Meadow Jumping Mouse (*Zapus hudsonicus*)

The meadow jumping mouse is another very small hibernator, with a head and body length of 3 inches (7.5 cm) and a weight of ½ ounce (14 g). It has a yellowish-brown coat with a dark dorsal stripe, pinkish-buff sides and white underparts, and dark ears that are edged with white. It lives in Alaska and southern Canada and throughout eastern United States from the Atlantic Coast back to Wyoming, and south to Oklahoma and Georgia, where it favors moist areas such as grassy meadows, aspen woods, willow thickets, marshes, and thick vegetation near the water's edge. It is a solitary and primarily nocturnal animal, and when hopping it uses its tail for balance, but jumps away erratically when startled, making leaps of up to 3 feet (90 cm). It is a good swimmer and climber and when it cannot reach seeding grass heads it either climbs the stems or cuts them off at the base. The meadow jumping mouse hibernates in the north from the end of October to late April, in a burrow dug into well-drained soil, in which it makes a nest of shredded grass at least 3 feet 3 inches (1 m) below the surface. Farther south its nest may be above ground in a sheltered place such as a hollow log or beneath a thick clump of grass. It is an omnivore, its diet of seeds, nuts, and berries being supplemented with insects such as beetles, caterpillars, and grasshoppers. It lacks cheek pouches and does not store food, relying on fat storage for its winter energy, and prior to hibernation its body weight doubles with accumulated fat. For its long sleep this species rolls into a tight ball with its tail curled around it and its body temperature drops to 35.6°F (2°C). It is long-lived for such a small rodent, its four-year lifespan being almost double that of a similar-sized vole; this is likely due to the fact that half its life is spent in torpor. Snakes, hawks, foxes, coyotes, weasels, and skunks are its major predators.

Jerboas

The jerboas, of which there are about thirty-two species, are the ultimate saltatorial rodents, with long tails, hind feet that are four times longer than their front feet, and the ability to bound bipedally like kangaroos in flight, covering up to 10 feet (3 m) in each bound and using their long tails for balance and as a tripod when standing. They are adapted for life in arid regions, which experience very hot summers and quite cold winters, and most species therefore have two periods of torpidity each year. They are nocturnal and spend the day sheltering from the sun and from hawks in their extensive burrow systems. A tuft of bristles around the ear opening prevents sand entering, and bristles on the feet aid walking on loose sand. They use their forelimbs for burrowing and their hindlimbs for kicking the sand away, and as they also push sand with their heads, the bone around the orbital cavities is thickened to protect their eyes; a fold of skin can be brought forward to cover the nostrils. Jerboas live in the southern Palaearctic Zoogeographic Region, including the Sahara, the Arabian Peninsula, and across central Asia to the Gobi.

Torpidity in the jerboas varies according to their locality. The three-toed jerboa (*Stylodipus telum*), which lives in the drier regions of central Asia from the Ukraine to Mongolia, hibernates from the end of September to mid-March. The four-toed jerboa (*Allactaga euphratica*) of the deserts of Asia Minor, Iran, and Afghanistan has a shorter sleep from late October to the end of February, and some populations may not hibernate at all. The lesser five-toed jerboa (*Alactagulus pumilii*) of central Asia hibernates from November to March. In both the five-toed dwarf jerboa (*Cardioranius paradoxus*) of the deserts of Mongolia and Northern China, and the fat-tailed jerboa (*Pygeretemus platyrus*) of central Asia, the tail thickens considerably with fat stored for the winter.

Desert Jerboa (*Jaculus jaculus*)

Desert jerboas are natives of North Africa and eastward through Asia Minor and the Arabian Peninsula to southwestern Iran, where they live in sandy regions, salt flats, rocky valleys, and dry grassland. They are one of the smaller species, weighing about 2 ounces (56 g), with a head and body length of 6 inches (15 cm) and a 9-inch-(22 cm) long tufted tail. Their color matches their habitat, being pale to dark sandy-buff above, with a white stripe on the hip and white underparts. They have large eyes and ears for good night vision and hearing in keeping with their nocturnal habits, and they can to leap up to 10 feet (3 m) horizontally and up to 3 feet 3 inches (1 m) high. They are nocturnal burrowers that dig counterclockwise spiral tunnels in sandy soil to prevent their collapse and enlarge the end for their nest and a storage chamber. In many regions children spend hours trying to catch them, after following their tracks in the sand and then probing their holes with sticks to waken the sleeping animal and flush it out of its bolt hole, when it is caught, roasted, and eaten. The desert jerboa is a solitary animal, and although its natural history is not well known it is believed to hibernate during the colder months, as individuals in a torpid-like condition have been dug out of their burrows in winter. They also estivate during long hot periods in midsummer. Like

Desert Jerboas   *Like the kangaroos they resemble, desert jerboas of North Africa and the Middle East are saltatorial and bound along on their sturdy hind legs, with their tail held out straight for balance. They may hibernate, but they certainly practice short-term hypothermia, becoming quite torpid during very cold weather.*
Photo: Friedrich Wilhelm Kuhnert, Brehms Tierleben

the kangaroo rats the jerboa does not drink and obtains metabolic water from its diet, which consists mainly of seeds, shoots, and roots.

Rough-legged Jerboa (*Dipus sagitta*)

Also known as the northern three-toed Jerboa, this species occurs in central Asia, from the Caucasus to Mongolia, where it favors sandy regions with scattered trees and shrubs. An attractive little animal, its upperparts are blackish-orange in winter and sandy-buff in summer, with a white hip stripe and a black-and-white tail tuft. It has a head and body length of 5 inches (13 cm) and its tail is 6½ inches (16 cm) long. Solitary and nocturnal, most populations of this species hibernate from November to March, surviving on the fat stored in their bodies in the fall. In some regions, however, especially Uzbekistan's Kyzyl Kum Desert, the period of torpor is reduced to just two months, and they have been seen out in midwinter on milder days. The rough-legged jerboa digs an extensive burrow system in sandy hillocks, which includes temporary ones for summer use, for daytime sleeping, and for quick access at night when they are threatened. A much deeper tunnel, down to 3 feet (90 cm) and with a chamber at the end and several emergency exits, is used for raising the family; and a deeper one still, down to 5 feet (1.5 m), is used for hibernation. This species prefers green foodstuffs such as the fleshy parts of plants, their stems, leaves, flowers, and fruit, but also eats seeds and insects.

Notes

1. A complete reorganization of the order *Rodentia* was recently proposed, with only two suborders—*Sciurognathi* and *Histricognathi*—based upon the shape of their lower mandible, but this arrangement has not been totally accepted, and the traditional classification is followed here.

2. Several frogs and young painted turtles also practice supercooling to avoid being frozen.

3. The flying squirrel's unrelated marsupial counterpart in Australia, which has evolved a similar way of life and therefore a similar form, is actually called the "sugar glider."

4. Wyoming pocket mouse (*Perognathus fasciatus*), Merriam's pocket mouse (*P. merriami*), silky pocket mouse (*P. flavus*), and the little pocket mouse (*P. longimembris*).

5. *Zapus hudsonicus*, *Zapus princeps*, and *Zapus trinitatus*.

6. Several flying mammals (bats) such as the pipistrelles and little brown bats weigh 1 g less than the southern birch mouse.

7 Light Sleepers

Carnivory, or the eating of animal flesh, whether of other mammals, fish, reptiles, or amphibians, occurs in most mammalian orders, but is most common in the order *Carnivora*, which contains the carnivores or meat-eating animals. This order is divided into the suborders *Arctoidea*, whose members are the dogs, bears, and bear-like carnivores, and *Aeluroidea*, which contains the cats and cat-like species. Hibernation is an uncommon behavioral trait in these animals; only nine species out of a total of 240 have a long winter sleep, and they are all members of the *Arctoidea*. Estivation is even rarer, with only the Asiatic black bear possibly estivating to escape the midsummer heat of the deserts of Pakistan and Iran.

As in all other hibernators the long winter sleep of the carnivores is a seasonal response to avoid inclement weather and the lack of food, and in all species it has two aspects in common. Storing food externally to eat during periodic arousal is unknown in the carnivores,[1] and without exception they store fat internally to provide the energy they need for hibernation. Second, their sleep is less profound than that of most mammals that become dormant in winter. Their body temperature does not drop significantly and they therefore sleep lightly, in a state of awareness, which is the major characteristic of their hibernation. Consequently, the carnivores are not usually considered hibernators, which have been defined as animals whose temperature drops drastically during their winter sleep. The carnivore's long sleep has confusingly been called torpidity, dormancy, winter sleep, and other names, rather than hibernation. However, the bears are in fact very efficient hibernators, perhaps more so than some of the so-called "true" hibernators, as they can sleep for up to seven months without eating, drinking, defecating, or urinating, whereas accepted hibernators like chipmunks and hamsters awaken regularly to raise their body temperature almost back to normal, to urinate, and to feed from their stores. Bears are unable to derive sufficient nutrients from the fibrous plant food available to them in winter,[2] and are not adapted to chase large animal prey through the snow like the wolves, so the only recourse for the terrestrial northern bears is to hibernate. As they also need to protect their cubs[3] during their first two winters, they evolved into very efficient hibernators, which actually sleep much longer than many so-called true hibernators. The winter sleep of the other carnivores is very similar to the bears' dormancy, although perhaps less intense, and none give birth while hibernating.

The hibernating carnivores awaken periodically and, excluding the bears nursing their cubs, may even go outside on mild days. Hibernating captive grizzly bears aroused daily for a few minutes to stretch and move around. The skunks and raccoons, which huddle together in family groups to keep warm, have only a small drop in body temperature and are awakened when the outside temperature rises. However, periodic arousals are not a negative factor for hibernation, as it is now known that the most profound hibernators, including even the alpine marmot and Arctic ground squirrel (which store fat and do not need to arouse to eat) all awaken periodically and raise their body temperature almost back to normal.

Storing food externally is typical only of the small omnivorous and herbivorous hibernators, and marmots are the largest species to do this, although generally they store fat internally for their long sleep. Carnivores use only fat for their hibernation energy, and do not normally draw from their muscles or organ tissues, other than a small amount by pregnant bears for the initial development of their cubs which are then born prematurely. It has been estimated that a female polar bear must gain 440 pounds (200 kg) of fat prior to entering her winter den, and the layer of fat on the hips of European brown bears prior to hibernation was 6 inches (15 cm) thick.

There are two major reasons for the higher temperatures of the bears during hibernation. First, they give birth and raise their young during their long sleep, the only animals to do so, and for this a warm body is essential in a cold den. Then there is the question of the cost-effectiveness of hibernating. Excluding the skunks and the raccoon dog, the hibernating carnivores are large animals, especially the bears, which have a thick insulating pelt and layer of fat and a much lower surface-to-body mass ratio than smaller hibernators, so they lose body heat much more slowly. Yet they cannot allow their temperature to drop too low, as the energy needed to raise it during their periodic awakenings would be more than the savings made by sleeping at a lower body temperature. The cost factor obviously applies to the bears because of their huge body weight, but also probably to the raccoon and the badgers, all of which weigh 22 pounds (10 kg) or more. The smaller carnivores often sleep communally to benefit from both their companions' body warmth and the general effect this has on the den temperature.

■ BEARS

Hibernation

Except in the southern parts of their range, all four species[4] of northern bears spend their winters asleep in a den, their temperature, respiration, and heartbeat all lower than normal, but higher than the profound hibernators, whose metabolism slows significantly and whose temperatures are close to freezing. The black bear's body temperature drops only to 88°F (31°C) from its normal 100°F (37.8°C), and although it sleeps soundly it can usually awaken quickly. Their heart rate drops from fifty to ten beats per minute, and their rate of breathing slows by half resulting in considerably reduced oxygen consumption. Consequently, their metabolism is reduced by perhaps 50 percent which is essential if their stored fat is to last all winter.

There are two reasons why bears hibernate—to protect their cubs and because their food is unavailable. Both reasons apply to the brown bears, black bears, and Asiatic black bears, whereas the female polar bear hibernates solely for the protection of her cubs; the males and nonbreeding females remain out on the ice all winter and find sufficient food. Bears are generally very hardy animals and they hibernate not just to avoid the cold but also because of the effect low temperatures have on their lifestyle.

There are several reasons why the bears' metabolism does not drop as low as that of the ground squirrels or hedgehogs, which are cold to the touch and unresponsive when hibernating. Their higher body temperature is necessary as they nourish the embryos, although briefly, and give birth and nurse their young while in hibernation. An animal in a deep torpid state, with its body temperature close to freezing and its metabolism and therefore its body functions barely operating, could not give birth, produce fat-rich milk for its young, or provide the warmth needed in a cold den to protect the tiny young, which in most bears are born naked. Stored fat is their only source of food for several months and consequently there is little accumulation of nitrogenous wastes. They neither urinate nor defecate during hibernation, and a plug called the tappen forms in the rectum. With their higher body temperature they sleep lightly and normally awaken quickly when disturbed, even by voices outside their dens, and are immediately in a physical condition to respond to potential threats, for their own protection and that of their young. This is not always the case, however. Biologist Lynn Rogers described falling into a black bear's den onto the female and her cub, but the female did not arouse for eight minutes despite the cub's cries. Bears move about during their hibernation, as much as possible in their tiny dens, and they periodically shiver violently to maintain their temperature at the lowered set point, which also helps to maintain muscle tone.

Physiology

The physiology of the bears during their long sleep has greater potential application to human medicine than any other group of hibernating animals. Nonhibernating mammals, including humans, need physical activity to maintain normal physiological functions, and inactivity results in dysfunction of the organs, organ systems, bones, and tissues. However, bears that are inactive for long periods and do not eat, drink, defecate, or urinate are not adversely affected. Despite being cramped in a small den for several months they avoid the problems of bone degeneration and muscle cramping that afflict immobilized humans, and their bones are as strong when they emerge from hibernation as when they started. The bones of humans immobilized for much shorter periods would have seriously degenerated, yet without exercise or weight-bearing movement, the bears can somehow maintain their blood calcium at the correct levels to avoid such deterioration. Bears also retain most of their muscle strength while hibernating, whereas immobilized people lose 0.7 percent of their strength every day, eventually leading to atrophy of the muscles.

Two factors are mainly responsible for maintaining the strength and tone of their muscles during hibernation—conserving protein and exercising. Bears "exercise" through massive shivering and muscle contractions which also raise their body temperature. Instruments charting their temperature show that these episodes may occur four times daily. They do not urinate during hibernation yet are not poisoned by urea, the major component of urine, resulting from tissue breakdown. They recycle the urea, converting its nitrogen into amino acids, which maintain their muscles and organ tissue during their long hibernation.

Another aspect of the bear's long sleep, of great interest to researchers hoping to help people suffering from kidney failure, is the fact that bears do not drink during hibernation yet they do not become dehydrated. Metabolizing fat to provide energy and water, they can maintain their body water balance for several months and can reduce the urine entering their kidneys by 95 percent while sleeping. A further side effect of the bear's metabolization of body fat for energy while hibernating is the doubling of its summer cholesterol level, yet it does not suffer the usual human afflictions—such as hardening of the arteries and cholesterol gall stone formation—resulting from high cholesterol. Another interesting and potentially valuable aspect of the hibernating bear's physiology is its ability to produce ursodeoxycholic acid, which has dissolved gallstones in humans.

Reproduction

The major advantage of the bear's den acting as a nursery is the protective start in life which it gives the cubs. Bear cubs raised in the den avoid their first winter outdoors and emerge ready to take advantage of the new season; they then have all summer to grow and store fat for the next winter, which they spend with their mother again, in the hibernation den.

Birth and raising during hibernation is the best possible arrangement considering the hostile climate the northern bears occupy, and to achieve this more secure start for their cubs they practice delayed implantation. Females are mated in spring or early summer, but the fertilized egg remains dormant for several months, and does not become implanted in the uterus wall and begin its development until it receives "a signal" to do so, the stimulus being readiness for hibernation. The eggs are absorbed if the female lacks sufficient body reserves to complete her hibernation. The total gestation period including the delayed implantation averages 230 days for all the northern bears, whereas the actual period of embryonic development is only about two months. The alternatives to this behavior are certainly not as acceptable. If bears mated in the fall like many northern animals, and still involved delayed implantation, after their 230-day total gestation period they would give birth in early summer, and the young would emerge from the den to find winter had begun. Mating in spring after emergence from hibernation, and allowing just two months for their gestation period and three months until they were ambulatory and emerged from the den, would also have them facing their first winter as tiny cubs.

As the hibernating mother bear does not eat during the growth of her embryos, it would be a great drain on her reserves to bring the cubs to full development in the uterus in the manner of the other placental mammals. Her cubs are therefore born prematurely, and are then nourished in a more efficient manner outside her body with her very rich milk, while protected in the den, a similar arrangement to the implacental marsupials (kangaroos and opossums), which raise their "premature" young externally in their pouches after a brief gestation period.

Bear cubs are quite underdeveloped at birth. Polar bear cubs are not hairless, but covered with fine, short hair at birth, whereas the other species are very sparsely haired or hairless, but they are very tiny compared to their mother. Newborn Asiatic black bear cubs weigh 8 ounces (250 g), black bear cubs weigh 10 ounces (285 g) at birth, grizzly bear cubs about 24 ounces (680 g), and polar bear cubs 22 ounces (625 g); but on their mothers' fat-rich milk they grow quickly.

Polar Bears   *Making their first appearance outside the den where they were raised, 3-month-old polar bear cubs are wary of the great expanse of snow and ice. Weighing only 22 ounces (625 g) at birth, during their mother's hibernation, they gained weight rapidly on her rich milk and now weigh about 26 pounds (12 kg).*
Photo: Courtesy USFWS, Steve Amstrup

They are fully furred when they leave the den at the age of three months in March or April, when polar bear and grizzly bear cubs weigh 26 pounds (12 kg), and black bear cubs 17 pounds (8 kg).

Although the female bear's metabolism is considerably reduced, she is still able to nourish her embryonic young, give birth, and then provide the cubs with milk. She could not do this if her metabolism approximated that of the hibernating ground squirrels, with her body temperature lowered almost to freezing, and with scarcely discernable respiration and heart rate. Bears are the only animals to lactate for several months without eating, while surviving on stored fat. Their milk is exceptionally rich in fat; polar bear milk is the highest with almost 35 percent milk fat. Brown bear milk contains 33 percent fat and black bear milk about 25 percent fat, and their milk is also high in protein and carbohydrates.

The Species

Polar Bear (Thalarctos maritimus)

The only semiaquatic bear, the polar bear is an animal of the Arctic regions, in North America on the northern coasts of Alaska and Canada, where it lives along the shoreline, on ice flows and pack ice and on the northern islands. It is an unmistakable animal, with its yellowish-white coat, large hindquarters, long neck, small head, and Roman nose. A mature male polar bear weighs about 1,435 pounds (650 kg), while females are much smaller at 660 pounds (300 kg). It is therefore slightly smaller than a male Kodiak bear, but is taller, reaching a length of 10 feet (3 m). The polar bear is well adapted to the Arctic weather and cold water, having a dense and thick underfur and an outer coat of long guard hairs that stick together when wet and form a waterproof layer. Its hairs are hollow and conduct ultraviolet light down to the black skin where the warmth is absorbed; it also has a 4-inch-(10 cm)-thick layer of blubber beneath the skin. The front paws are very broad for paddling, and the soles of its feet are covered with small papillae that give good adherence to ice. It is almost totally carnivorous, seals being the main prey, plus beluga whales and walruses, and carrion is also relished, especially from beached whales.

The polar bear ranges south with the pack ice in winter, sometimes reaching Newfoundland, Iceland, and the Pribilof Islands in the northern Pacific Ocean south of Alaska. There are permanent populations in Hudson's Bay and James Bay, at the same latitude as northern England, where much of the time is spent on land in the summer and fall when the absence of sea ice makes seal hunting impossible. During this lean time their metabolism slows and they use less energy, surviving on their stored fat in what has been called "walking hibernation." They raid garbage dumps like the one at Churchill, Manitoba, where the town was built on a traditional migration route, and the bears have become a nuisance. During the 1970s many females were sedated and helicoptered out of town to the barren grounds, while their cubs were shipped south to Winnipeg's Assiniboine Park Zoo, from where they were sent to zoos worldwide, excluding the United States, as the Marine

Mammal Act required imported polar bear cubs to be accompanied by their mothers, and exceptions were not made for babies prematurely but permanently separated from them. Once approaching endangered species status, the polar bear's numbers are now stable or increasing, due to protection from hunting, except for native "subsistence" hunting for meat (although the liver must not be eaten due to its toxic concentrations of Vitamin A); but the increasing development of the Arctic is now again raising concern for its long-term future.

Polar bears are solitary animals; they meet briefly to mate in spring or early summer, and when two or more are seen together for long periods they are undoubtedly a mother and her cubs. A fast runner, it also swims powerfully with its head and neck out of the water. The mating season is the month of April but delayed implantation prevents the development of the embryo until September, when the fertilized eggs attach to the wall of the uterus when the animal's condition is most favorable for their development—when fat has been deposited for the winter hibernation. The actual period of embryonic development of about sixty days then begins, and from mating to birth the total gestation period averages 230 days.

Pregnant polar bears hibernate on land and dig a den in a bank or deep snowdrift on a slope near sea ice in August or September. A typical den consists of an entrance tunnel about 6 feet 6 inches (2 m) long leading to the sleeping chamber, which is not large for the size of the bear, usually only about 5 feet (1.5 m) in diameter and 3 feet 3 inches (1 m) high. The entrance tunnel slopes upward so the chamber is higher and warm air is trapped, and the chamber temperature is usually just below freezing, whereas it may be −40°F (−40°C) outside. Males and nonpregnant females generally do not hibernate, but stay out on the sea ice hunting all winter, and den only during very severe weather. The female polar bear's heartbeat drops from 70 to 10 per minute during hibernation, and her temperature drops from 98.5°F (37°C) to 88°F (31°C), which is certainly high for a "dormant" animal but necessary for pregnancy, birth, and nursing. The cubs, normally two, are born in the den between the end of November and early January. They are very small for the mother's size, weighing about 22 ounces (625 g), and are blind and covered with short, fine hair. The mother and her cubs stay in the den for three months, and then emerge into the snow, at which time the cubs weigh about 26 pounds (12 kg). The mother bear loses up to 300 pounds (136 kg) while in hibernation. There is high mortality of the cubs in their early months (mainly from starvation and predation by male bears) but growth is rapid for the survivors, and when they hibernate with their mother for their first full winter they weigh about 100 pounds (45 kg).

Brown Bear

Primarily a mountain animal, where it inhabits the upper forest zone and alpine clearings, the brown bear is one of the most widely distributed of all land mammals. In the New World it occurs in the mountains of western North America and northern Mexico, on the barren tundra, along the Alaskan coast, and on fog-shrouded islands off the coast of Alaska, and until recently on the prairies. In the Old World it ranges from the Pyrenees westward across Eurasia to Kamchatka and

Hokkaido, and also occurs in the mountains of North Africa. It is distinguished from the black bear by its size, its distinctive shoulder hump, and longer coat. Most bears are dark brown but there is some individual variation, plus subspecific color differences over the vast range, and there is considerable variation in size. The subspecies include the grizzly bear and Kodiak bear in North America and the European bear and Syrian bear in Eurasia.

Brown bears everywhere have been affected by development and hunting, and are now extinct or rare in many parts of their original range. They have been extinct on the prairies for almost two centuries and are rare on the Arctic Barren Grounds and in the contiguous United States. They are also very rare in Morocco's Atlas Mountains and on Japan's northern-most large island of Hokkaido; are virtually extinct in the Pyrenees and Cantabrian Mountains, and are much reduced throughout the rest of Europe. Only in central-eastern Asia and on the Alaskan islands are they still plentiful.

Solitary animals that mate and then go their own way, their only social bond is that between a mother and her cubs. Their hibernation varies in keeping with the local environment, from at least six months in the coldest regions to none at all in northern Mexico. They must lay down sufficient fat for the hibernation period, otherwise their chances of survival are slim, as arousing and finding food in winter is virtually impossible; it is believed they do not even begin their sleep until they have stored enough fat. They seek dry places to hibernate, such as a natural hollow between rocks, a cave, beneath the root mass of a fallen tree, or the hollow base of a standing tree. The entrance to their den is often covered with snow, completely hiding it from view.

Brown bears are omnivores that eat a wide range of plant and animal life, but they are also fast and agile predators. In summer they can run down caribou, elk, and moose, and many grizzlies acquire the fat for their winter survival by gorging on salmon "running" upriver to spawn. On the tree-less Arctic tundra where vegetation forms the bulk of their diet, Barren Grounds grizzlies gain calories for the winter and replenish their reserves when they emerge from hibernation by digging ground squirrels out of their burrows; and during six weeks' observation one was seen to capture 350 squirrels, which hibernate in shallow burrows due to the permafrost.

Grizzly Bear (*Ursus arctos horribilis*)

The grizzly bear is the familiar subspecies or race of the brown bear in western North America, occurring on the northern tundra of Alaska and northwestern Canada, and southward through the Rockies at high elevations to northern Mexico. Although smaller than the Kodiak bear it is still a very large animal, with adult males weighing up to 880 pounds (400 kg). It is a distinctive race that varies from pale yellowish-brown to dark brown, with the white-tipped hairs on the back producing a frosted or grizzled appearance, and giving rise to the names "grizzly" and "silver-tip." It is dish-faced in profile with a noticeable hump on the shoulders.

The grizzly bear is a solitary animal, except during its short mating season and when females are accompanied by their cubs. It is omnivorous and eats virtually anything of plant or animal origin, including grass and other fresh vegetation, berries, and rodents, and gorges on salmon when they are spawning. It eats carrion, kills black bears, and usurps wolves from their kills; despite its bulk it is very quick and agile over short distances and can outrun juvenile elk. The grizzly bear is considered the most dangerous North American carnivore, mainly as a result of its increased contact with humans, who have encroached on its once remote habitat. It digs its own den, where it makes a bed of grass and leaves, and often uses the same one for several winters, changing the bedding each year. It begins its hibernation in October, with lowered heart rate and respiration but only a small drop in temperature, and therefore arouses quickly when disturbed. The cubs, usually two, are born in the hibernation den and emerge with their mother in April.

Grizzly Bear   *The carnivore hibernators all rely entirely on stored body fat for the energy needed for their winter dormancy. The grizzly bear lays down a thick layer of fat, especially on its hips, gained largely from gorging on migrating salmon in the fall.*
Photo: Courtesy Harcourt Index

Kodiak Bear (*Ursus arctos middendorffi*)

The Kodiak bear is the world's largest bear, and therefore the largest living land carnivore, much larger than the grizzly bear and weighing up to 1,545 pounds (700 kg). A lowland animal, it lives mostly at sea level and along the coastline of the islands that comprise the Kodiak Archipelago off the southern coast of Alaska.

It is normally dark chocolate-brown, with the occasional golden-brown individual. It mates in May or June, but delayed implantation of the fertilized eggs in the wall of the uterus interrupts their development until the fall. Pregnant females begin their hibernation in late October, and the males enter their dens soon afterward and emerge in April before the females and their cubs. The tiny cubs, up to four but usually two, are born in the den in January or February, and at birth weigh about 24 ounces (680 g) and are naked and blind. They grow rapidly on their mother's rich milk and weigh 26 pounds (12 kg) when they emerge with her in April or May. They stay with their mother for three years, but there is very high mortality rate in bear cubs, especially due to cannibalism by mature males. Living close to the sea gives the Kodiak bear feeding opportunities denied other brown bears, and it eats carrion—beached whales and fur seal carcases—plus seaweed in addition to the other more typical bear foods such as vegetation, berries, ground squirrels, and salmon.

European Brown Bear (*Ursus a. arctos*)

The brown bear of Europe is now a very rare animal, which survives mainly in small and isolated populations. In central and western Europe only a few remain in Spain's Cantabrian Mountains, in the Pyrenees, the Alps of France and Italy, and in central Italy's Abruzzo Mountains. In western Europe it can still be found in northern Scandinavia, in the Carpathian Mountains of Romania, and in the Balkan Mountains of neighboring Bulgaria. It has also survived in the mountains of northern Greece from where it has recently extended its range to the south almost to the Gulf of Corinth. Romania, where bears were protected for the dictator's personal hunting during the communist regime, harbors the largest population, which must now be culled annually to prevent overstocking. The European brown bear resembles the grizzly bear, but is slightly smaller, with adult males weighing up to 825 pounds (375 kg), but it has very similar biology. Its diet consists of vegetation, colonial insects, fruit, berries, rodents, deer, domestic livestock, carrion, and garbage. Mating occurs in May and June, with delayed implantation interrupting the development of the fertilized eggs until the fall. Bears enter hibernation at this time, when the fat on their hips may be 6 inches (15 cm) thick. The cubs are born in the den in December or January and emerge with their mother in March or April.

American Black Bear (*Ursus americanus*)

The black bear is the common bear of North America, occurring in Alaska and Canada, and south through the western mountains to California and northern Mexico, avoiding the Great Basin, and in the east southward through the forests to Alabama. It also occurs along the Gulf Coast and in Florida. Numerous races or subspecies are recognized over this large range, and their choice of habitat is also wide-ranging, including forests, tundra, prairie river valleys, dry mountain scrub, and the swamps of the southeastern United States. It is still plentiful in many parts of its range, where it is considered a game animal with a regular hunting season, when thousands are shot annually; but in other regions, notably Mexico, Florida, and the Mississippi Valley, it is now rare. It is a solitary animal, and a male's range usually overlaps those of several females, but they avoid close contact even when congregating at a garbage dump.

The black bear is smaller than the brown bear, and stands 3 feet (90 cm) high at the shoulder, with the average weight for males being 507 pounds (230 kg) and for females 275 pounds (125 kg). It lacks the brown bear's shoulder hump; it has a shorter coat and shorter claws, and a long and sloping head from ears to nose, whereas the brown bear has a slightly upturned muzzle. Black bears are not always black; color phases are quite common—especially brown, blond, cinnamon, and creamy-white—often with more than one mutation in the same litter, and the darker forms usually have a white patch on the chest. The blue or glacier bear is a dark bluish-black race with numerous grayish hairs, which lives on the coast of southern Alaska.

The black bear has been called the perfect omnivore, for despite its large size it is specialized to eat small, easily obtained and digested items, such as fresh grass and leaves, roots, tubers, berries, acorns, nuts, honey, colonial insects such as wood ant and bee larvae, crayfish, frogs, rodents, deer fawns, garbage, and carrion, and it usurps cougars' kills. It is an agile tree climber and a good swimmer, and is most active at night.

Mating occurs from mid-June to mid-July, but the gestation period of 230 days is interrupted by delayed implantation, and the actual period of embryonic development is only sixty days, which begins in late November. The cubs, which average 10 ounces (285 g) in weight are born in late January in the hibernation den. Unlike the polar bear, both male and female black bears hibernate, although separately, and may increase their weight by half prior to their long sleep, with carbohydrate-rich berries providing much of this fat reserve. The den is usually quite small and cramped, requiring the bear to curl tightly, which conserves energy. Small dens may be thought a means of retaining heat, but many are actually open to the elements, and with their lowered body temperature and fat-insulated bodies, plus their curled position to protect their lightly furred belly and the nursing cubs, bears generally do not radiate sufficient heat to warm the den. Therefore, contrary to popular belief, their body warmth does not melt the snow inside, above, or around the den, where the temperature may be little warmer than the ambient temperature. A hole in a bank which they have dug themselves is a typical den, or they may dig beneath the roots of a fallen tree or sleep in the hollow base of a large conifer, and may use the same site for many years. They usually make a nest by pulling in pine branches or piles of grass, but sometimes just lie on the bare earth.

Black bears are very efficient hibernators. Despite the coolness of the den, their thick layer of fat, the insulation value of their dense fur, and their low surface-to-body mass ratio all help to retain the heat generated by their store of fat, and they do not lose heat quickly. When hibernating, their body temperature drops from

100°F (38°C) to 88°F (31°C), their metabolic rate is depressed, and respiration slows. Their heartbeat drops from about 45 per minute in normal sleep to 8 per minute during hibernation, and they usually sleep lightly and can rarely be closely approached without awakening. Although their metabolic rate may be reduced by half when they are sleeping, nursing females still lose at least 40 percent of their body weight during hibernation, due in large part to the production of fat-rich milk for their cubs.

The hibernation period of the black bear varies according to local conditions. In southern Manitoba and northern Minnesota they usually sleep from October to March. In Washington's milder climate their average winter sleep lasts 126 days and in Louisiana about 80 days, whereas in Florida they do not sleep at all. Where they do hibernate in the milder latitudes they may awaken on warmer days and may even leave the den briefly. During their long sleep they neither defecate nor urinate, and recycle their waste products, breaking down the urea produced from fat metabolism and making protein from the resulting nitrogen, allowing them to replenish their muscles and tissues. They awaken to give birth, usually to two cubs, which are naked and blind, and then return to sleep, with their cubs safely tucked into the warmth of their sparsely furred stomach, with access to the nipples. The mother arouses periodically to lick her cubs, but otherwise continues hibernating, and they all leave the den when the cubs are three months old and weigh about 17 pounds (8 kg). The cubs den together with their mother for their second winter, and females therefore breed in alternate years and sometimes only every three years.

Asiatic or Himalayan Black Bear (*Selenarctos thibetanus*)

The Asiatic black bear ranges across southern and central Asia from Iran to the Pacific Coast, north to southern Siberia and south to Thailand. It also occurs on the islands of Taiwan, Hainan, and Japan. Its preferred habitat is deciduous forests at high elevations in summer, from where it moves down to sheltered valleys in winter, but it also lives in the semiarid regions of Pakistan and Iran. It is a slightly smaller animal than the American back bear, with a longer coat, especially on the neck and shoulders and giving the appearance of a mane, and with larger and more rounded ears. Adult males reach a maximum weight of 485 pounds (220 kg) and females 250 pounds (114 kg). Although black is the most common color, brown individuals also occur, but all have a white crescent-shaped marking on the chest.

This bear is mainly nocturnal and sleeps most of the day in a cave, in the base of a hollow tree, or in a large nest which it builds high in a tree. It is completely omnivorous, with a diet typical of the other terrestrial bears, which includes a wide range of plant and animal foods. A very agile animal, it climbs palm trees to gather nuts and to eat the growing hearts, and with its powerful limbs and long claws tears off bark to reach the sap wood and rips open termites' nests to eat termites and their larvae. It is a very aggressive animal, which has frequently killed people, due mainly to loss of habitat to farming, and is now raiding croplands for grain, vegetables, and domestic stock. Consequently, it is resented and persecuted, and is locally endangered.

The Asiatic black bear hibernates in Siberia from October to March, but not at all in Burma, Malaysia, or Thailand. In the hot deserts of Iran and Pakistan it may estivate briefly in midsummer. Its reproduction hinges upon locale, climate, and the need to hibernate. In Siberia it mates in midsummer and following delayed implantation gives birth in the hibernation den in January or February, after an actual embryonic development period of sixty days, which is typical of all bears. In southern Pakistan, where it does not hibernate, mating occurs in October and the cubs are born in February, blind and hairless and weighing only 8 ounces (225 g). They stay with their mother for three years.

Other Carnivores

The other carnivores that hibernate also belong, like the bears, to the suborder *Arctoidea* and include the badgers and skunks of the family *Mustelidae*, the raccoons *Procyonidae*, and the raccoon dog, which is the only member of the dog family *Canidae* to hibernate. No members of the other carnivore suborder *Aeluroidea* (the cats and cat-like carnivores) hibernate, nor are they known to practice daily hypothermia. Like the bears these carnivores do not sleep with the deep intensity of the ground squirrels or bats. They are light sleepers, an indication that their temperature is not significantly lowered, and they are therefore not in deep torpor. The slight drop in temperature is still sufficient to lower their energy requirements,[5] which together with their denning behavior enables them to survive periods of inactivity in cold weather. These periods may last for almost five months in species living in the far north, during which time their energy is provided entirely by stored fat since they neither eat nor drink while inactive. Like the bears their sleep has been called many names other than hibernation, simply because they did not fit the old definition of that term, which implied sleeping with a drastically lowered temperature. All of these hibernating carnivores are omnivorous; they are also nocturnal and therefore do not need to estivate to escape high daytime temperatures.

Some of the Species

Raccoon (*Procyon lotor*)

With its black mask and ringed tail, the raccoon is the most famous and easily recognized North American mammal. It occurs throughout the United States except for some western desert regions, and extends south into Mexico and north into southern Canada. It was introduced into Europe in the 1920s, and is now established in Germany, Poland, and Belarus. The raccoon is mainly nocturnal and is only occasionally seen in daylight, but pet animals usually change their habits. It is often kept as a pet but may outgrow its docility, and with its powerful jaws and often irascible temperament when old, can give a serious bite. It is a good climber with great forearm mobility and dexterous hands, and is very hardy, able to withstand low temperatures; but it uses burrows, tree cavities, or attics and lofts to shelter from severe cold.

Adult male raccoons weigh up to 27 pounds (12 kg), which can increase by 50 percent when they store fat for their long sleep. Their hibernation period varies considerably according to locale. Heavy snowfall and temperatures below 20°F (−7.3°C) usually encourage them to commence hibernation, and when the temperature rises close to freezing they emerge to search for food. At the far north of their range, in the harsh midcontinental climate of the Canadian prairies, they stay in their dens for most of the severe winter—from late November until late February—whereas in Arizona and Mexico they are usually active all year. Like the bears they are light sleepers, and their body temperature drops only a few degrees, to 97°F (36°C), from their normal 102°F (39°C), so their metabolic rate hardly changes, and there is little or no change to their heart rate or respiration. Their habit of choosing a very sheltered site for denning and of sleeping communally— over twenty animals have been found in winter dens—undoubtedly helps to conserve energy and allows them to remain inactive for long periods when their metabolism functions almost normally. Raccoons are omnivores, both predators of small mammals and birds and eaters of carrion and human garbage, and they raid gardens and fields for grain, vegetables, and fruit. Crayfish and frogs are favorite

Raccoon *A raccoon peers out of its snowbound den on a mild day in midwinter. Typical of the hibernating carnivores its metabolism does not drop significantly even during severe weather, but it does remain asleep for several months in the far north of its range.*
Photo: Clive Roots

foods, and searching underwater for them has developed into the habit of dunking other foods, when they are mistakenly thought to be washing them.

Raccoon Dog (*Nyctereutes procyonoides*)

The raccoon dog is the only member of the dog family *Canidae* to hibernate, but only does so in the northern parts of its range. It is a small animal, with a head and body length of 23 inches (60 cm), a tail about 7 inches (18 cm) long, and an average weight of 15 pounds (7 kg). Although a true dog, it resembles a raccoon, with a thick body, bushy tail, and a grizzled-gray coat with dark areas around the eyes. It evolved in the Far East and its natural range until early last century was from southeastern Siberia southward through Mongolia, Korea, and China into northern Vietnam, and on several of the Japanese Islands. However, its fur was believed to have commercial value and thousands were liberated into European Russia, west of the Ural Mountains to develop a fur trade there. From there it spread rapidly to Scandinavia and France, occupying every country on the way. It is now considered a noxious alien throughout Europe, as it preys upon game birds such as partridge, quail, and pheasants. It is no longer trapped as its pelt is of poor quality due to the milder winters, and is unfashionable and rarely used nowadays anyway. Also, it is a vector of rabies and was blamed for an epizootic in Finland in 1989.

The raccoon dog prefers deciduous forests, reed beds, and grassland in the vicinity of streams and lakes where there is dense underbrush. Primarily nocturnal, for its daytime sleep and for hibernation it digs its own holes, but also uses the burrows of foxes and badgers, plus natural rock crevices and hollow trees. It is omnivorous and has a very wide diet of animal and plant foods, including small mammals, birds' eggs and nestlings, fish, crustaceans and frogs; and the abundance of food in the fall—especially cultivated cereals, fruit, and berries—allows it to increase its body weight by 50 percent with accumulated fat. Life in the far northern parts of its range is very hard for this animal, and only those that have laid down stores can safely hibernate. In Siberia, Mongolia, northern China, and on Hokkaido, the mated pair hibernates with their current litter from late October to the end of March. They may appear on mild days and attempt to find food, but such days are rare in that part of the world, whereas in the warmer climate of southern China and Vietnam they do not hibernate. In parts of its new range—in Azerbaijan, Belarus, Ukraine, and in the Caucasus Mountains—the raccoon dog sleeps while there is snow on the ground, which is normally for almost five months. In Finland, the northern limit of its new distribution extends almost to the Arctic Circle where it is subfreezing for several months, and it hibernates from early November to March. During hibernation its metabolism is reduced, although its body temperature is not significantly lowered, so it is a light sleeper.

Eurasian Badger (*Meles meles*)

The badger of the Old World is a grayish animal with dark belly and limbs and a white head with black stripes from its ears to its nose. It has a head and body length of 36 inches (92 cm) and a short tail, and adult males average 30 pounds (14 kg) in weight, which can almost double in late fall with accumulated fat. Its wide body and short and sturdy limbs give the animal a flattish appearance. This badger is a nocturnal resident of the British Isles, most of lowland Europe and Asia from the Atlantic to the Pacific coasts and Japan, and south to Asia Minor and Iran. It is an animal of forest and farmland—provided there is adjacent woodland—where it lives in large communal burrows called setts. These extensive burrow systems may be used for many years by several generations of badgers, who will drag grass down to make their nests and change their bedding each spring.

The Eurasian badger is an omnivore that eats a wide range of plants and animals. Earthworms are a major source of food, plus snails and insects, rodents, young rabbits, birds' eggs and nestlings, frogs, reptiles, acorns, berries, fruit, crops such as corn and potatoes, rhizomes, fungi, and carrion. It mates in midsummer, but delayed implantation of the fertilized eggs occurs and they lie dormant at the blastocyst stage for several months before development begins, so that the cubs are born at a favorable time of year. This is in February or March, after an actual period of embryonic development of seven weeks, so the cubs emerge from the den in late spring. The badger hibernates during cold weather and heavy snow, which in central Norway and Sweden and to the northern limit of the tree line in Siberia lasts from October to April, although it may appear on warm days. Like the other hibernating carnivores it is a light sleeper, for its body temperature drops only to about 94°F (34.5°C) from the normal 99°F (37.5°C), and its breathing and heart rate are only slightly reduced. However, even such a small reduction in its metabolism can reduce its energy requirements sufficiently to extend the life of its fat reserves, aided also by the protection of its underground nest and communal sleeping habits. Farther south, in the Caucasus, it hibernates only from November to March and in Kazakhstan for only three months each winter. In England badgers are generally inactive in their setts during December and January.

American Badger (*Taxidea taxus*)

The American badger has a grayish body and a dark face, with a white line through the eyes, a white patch below the ears, and a thin white stripe from nose to nape. It has the stocky body and short legs typical of the badgers, which makes them appear flattened dorsoventrally. Its habitat is the grasslands and farmlands of central and western North America from southern Canada to Mexico, where it lives socially for most of the year in a complex system of burrows and chambers. These are dug by the members of the group with their powerful front feet and may be 30 feet (9 m) long and 10 feet (3 m) deep, well below the frost line, even in the northern prairies. It is slightly smaller than the European badger and the maximum weight of a mature male is 24 pounds (11 kg) prior to accumulating fat for hibernation.

Badgers spend the day in their burrows, appearing at dusk to hunt rodents and young rabbits, which they do not chase but dig out of their burrows, plus reptiles, amphibians, carrion, and invertebrates—bee and wasp larvae, snails, and worms. They also eat a wide range of plant matter such as nuts, fruit, acorns, tubers, and

American Badger *On the northern prairies badgers sleep in their underground dens from November to March, dependent for their energy on stored fat. Their body temperature drops only a few degrees and they sleep lightly, usually denning communally to benefit from the combined warmth of the group.*
Photo: Courtesy Harcourt Index

mushrooms, and raid cultivated crops of grain, vegetables, and fruit. They are noted for their playfulness and their cleanliness—they dig holes for their feces and change their bedding regularly. They stay in their burrows when it is very cold or if there is deep snow, and may be inactive for several months in the north, living off their fat reserves.

The length of dormancy depends entirely upon the local climate, with animals in southern Manitoba and North Dakota sleeping in their dens from November to March when food is unavailable because the ground squirrels are also hibernating deep below the frozen soil, and practically all the waterbirds have migrated. Like the other light-sleeping carnivores, the badger's body temperature, heart rate, and respiration drop little during hibernation. In the milder latitudes of Kansas and Iowa they are inactive for two months, while in Texas they only become dormant in severe weather.

Striped Skunk (*Mephitis mephitis*)

The striped skunk is predominantly black, with a white nape that divides into a V and extends back to the base of its bushy black tail. When adult it averages

8 pounds (3.6 kg) in weight, prior to storing fat for the winter, which may add another 5 pounds (2.2 kg). It has a wide range throughout North America, from southern Yukon and the Northwest Territories in Canada through the United States into northern Mexico, where it lives in woodland, bush, and grassland, generally near water. Skunks have an excellent sense of smell and hearing, but are near-sighted and unable to focus clearly on objects more than 10 feet (3 m) away. The striped skunk is mainly nocturnal and is seldom seen during the day, when it hides under a pile of brush, logs, or rocks, in a hollow tree trunk, beneath buildings, or in a burrow, often one expropriated from a woodchuck. It crops grass and collects leaves for its day nest, which is also used for hibernation. Its diet includes berries, fruit, buds, grain, acorns, frogs, salamanders, insects, snakes, birds' eggs and their nestlings, small rodents, carrion, and garbage, and by the fall it appears very stocky due to its thick layer of fat. In North Dakota and Manitoba females and their young begin hibernation in October when the temperature drops to freezing, with the males following when the temperature reaches 14°F (−10°C) overnight. It does not emerge until March, having lived entirely off its fat reserves during this inactive period. Farther south skunks hibernate for progressively shorter periods; in Iowa they are inactive from early December to the end of February, and in Texas they may be seen all winter except on very cold days. Like the other carnivores skunks are light sleepers, and when dormant their metabolism is only slightly reduced, with their temperature dropping from 98.6°F (37°C) to 87.8°F (31°C), but this is sufficient to considerably reduce their energy use. Their habit of denning communally, at least several females and their young sleeping together, also helps to conserve energy.

Notes

1. During the active season badgers sometimes store food to eat later, but they do not store food for their inactive period.

2. Unlike the nonhibernating giant panda, which has access all winter to the bamboo that forms the bulk of its diet.

3. Which is why the marine and semiaquatic female polar bear hibernates, whereas males are usually active all winter.

4. The polar bear (pregnant females only), black bear, Himalayan black bear, and the brown bear (*Ursus arctos*), which has been subdivided into many species in the past but is considered a single species, with several races that include the grizzly bear, Syrian bear, and Kodiak bear.

5. Reduction of the body temperature by 18°F (10°C) is generally believed to halve the rate of energy consumption.

8 Bats at Rest

Bats are the only mammals that can truly fly, and they form the second-largest group after the rodents, their order *Chiroptera* containing almost 1,000 species. Their wings are very thin and elastic membranes, formed by two layers of skin which extend from the sides of their body and hind legs, enclosing the elongated bones of the forelimbs, the phalanges of the hands, and the nerves and blood vessels. Their sternum, like that of the flying birds, is keeled for the attachment of the breast muscles which power the wings on the downward stroke, but unlike birds the flight of bats involves both the wings and legs working in unison. Fossils prove that the bats of 50 million years ago looked very similar to those alive today, but the ancestors of those early bats have yet to be discovered. It seems likely that they evolved from insectivorous land mammals, probably becoming first adapted for gliding like the present-day "flying" squirrels and sugar gliders, and then progressing to the highly modified fliers of today.

Bats are not blind; they can all see and the tropical fruit-eaters have very good vision; many rely upon sight to find their food. However, the insectivorous species depend upon echolocation to maneuver and locate their prey. Their high-frequency sounds, which are beyond the range of hearing of the human ear, are emitted in flight through the nose or mouth and are reflected back as echoes. This system also depends upon exceptional hearing and rapid motor response, and the most distinctive general features of the bats, after their wings, are their strange nose appendages or "leafs" and their ears, which are very long in many species.

All bats are nocturnal and when they rest during the day most species seek shelter where the microclimate and darkness provide both security and a suitable environment, especially caves, mines, hollow trees, and within buildings, but some have more unusual preferences. The tomb bats (*Taphozous*) are so named because they prefer to roost in tombs, and in Egypt the mouse-tailed bats (*Rhinopoma microphyllum*) hibernate for several weeks in pyramids. Others roost in the open, some hanging partially hidden beneath leaves, others fully exposed to the sun and predators in communal tree roosts. Most species inhabit the warmer regions of the world, where food (fruit, nectar, insects, small animals) is available year-round and neither migration nor hibernation is necessary, but those living in northern temperate regions are strictly insectivorous and must avoid the cold months when their food is no longer available. At that time of year the northern members of the order *Chiroptera* are therefore totally absent or inactive,[1] having migrated or hibernated, more so than the members of any other order of northern temperate zone land mammals.

Hibernation rather than migration is the most prevalent behavior of the northern species, and bats that do this belong mainly to the families *Rhinolophidae* (horseshoe bats), *Hipposideridae* (Old World leaf-nosed bats), and to the very large family *Vespertilionidae*. The latter contains many small insect-eating species, especially the little brown bats (*Myotis*), the pipistrelles (*Pipistrellus*), noctules (*Nyctalus*), and the serotines (*Eptesicus*). They are the smallest hibernating mammals, with several species of *Myotis* bats weighing only ⅕ ounce (5.6 g), and the eastern pipistrelle (*Pipistrellus subflavus*) and the common pipistrelle (*P. pipistrellus*) being the smallest mammalian hibernators at just ⅙ ounce (4.8 g).

These northern bats are said to be heterothermic mammals, as their body temperature fluctuates in accordance with the ambient temperature, both on a daily basis and for much longer periods, and they only maintain their normal body temperature when they are active. When resting by day or hibernating their temperature drops close to that of their surroundings, with associated reduced breathing and heart rate; diurnal torpidity differing from hibernation only in its duration. As winter approaches and food supplies dwindle, they cannot keep warm and must avoid freezing temperatures. Successful hibernation depends upon the storage of sufficient fat reserves during the late summer and fall to provide the energy needed to fuel several months of sleep, and many bats gain fat reserves equal to their normal body weight prior to hibernation. This stored fat is not only to fuel their lowered metabolism during the long sleep but also to rewarm them during their natural periodic arousals. However, hibernating bats are easily disturbed, and awaken quickly, and the energy used during each complete arousal has been estimated at about 10 percent of their fat reserves, which affects their ability to survive the winter. Consequently, access to their hibernacula has been prevented in Europe and America to protect the sleeping bats.

After fat storage, finding a hibernacula with the correct temperature and humidity is the most important factor for their successful hibernation, and they generally use the same place for many generations. It must protect them from the elements, and have a fairly constant temperature, the preferred range being between 37°F (2.8°C) to 42°F (6.1°C), and the humidity must be high (66–95%) to prevent dehydration and their wing membranes drying and cracking. They sleep in buildings, attics, belfries, and tree holes, but underground provides the most suitable and stable conditions, and caves and mine tunnels and shafts are favorite sites, offering a stable microclimatic environment where they can safely sleep and conserve their small reserves of fat. Bats cannot withstand frost, and if their hibernacula is too warm the corresponding rise in their body temperature will use

Indiana Bats   *Most northern bats, which are all insectivorous, hibernate communally in caves where the humidity is high and the temperature remains above freezing. Millions may congregate clustered together for the winter at popular roost sites, which have been used traditionally by many generations of bats, like the Indiana bats above.*
Photo: Courtesy Indiana DNR, Scott Johnson

their fat stores too quickly and they would starve. Consequently, they often make long journeys to a traditional trustworthy hibernacula. Gray bats (*Myotis grisescens*) may fly 300 miles (482 km) from their summer habitat to their traditional wintering cave in Missouri. Favored caves attract enormous numbers of bats, and Kentucky's Mammoth Cave was once home to an estimated 10 million gray bats and Indiana bats (*Myotis sodalis*). They may also change their hibernacula in midwinter if the conditions become unsuitable.

Despite their need to hibernate there are numerous accounts of bats being observed flying at temperatures well below freezing, further proof that lack of food rather than inability to withstand cold is usually the main reason for torpidity. The small-footed myotis (*Myotis leibii*) of eastern North America has been seen flying when the temperature was only 15°F (−9°C), and a noctule bat (*Nyctalus noctula*) flew over the campus of the University of Moscow one December night when the temperature was 23°F (−5°C). Such unusual behavior is unlikely to be associated with feeding, as the improbability of finding insects then compared to the use of stored energy would not be cost-effective. It is known that hibernating concentrations of bats do not remain constant all winter, and they may awaken during their hibernation and change their site, flying a mile or more to a new one.

The northern bats breed prior to commencing hibernation, but most, if not all, practice delayed fertilization, in which the spermatozoa is stored in the female's uterus until spring when ovulation and fertilization occurs. The hibernation period of the most northerly ranging insectivorous bats, such as the little brown bat (*Myotis lucifugus*) and the eastern pipistrelle (*Pipistrellus subflavus*), usually extends from September to April. In some species, such as the Indiana bat, adult females enter hibernation several days earlier than the males, and then emerge in March or early April, the males leaving later in April. This is possibly associated with the commencement of their pregnancy and the need to feed quickly.

The metabolic rate of hibernating bats is very low, that of the pipistrelle being only 1 percent of its normal resting rate, compared to 5 percent for a marmot, illustrating that smaller mammals have the most effective energy reduction during hibernation. The body temperature of hibernating bats drops from the normal 99°F (37.5°C) to match that of their surroundings, usually between 33°F (0.5°C) and 42°F (5.5°C) and their bodies are cold to the touch. Their heartbeat drops to 25 per minute from about 400, and their breathing is imperceptible, so their oxygen consumption is therefore negligible. In fact, they may not breathe for long periods, which has been proved by noctule bats surviving immersion in water for sixteen minutes. However, some can withstand much lower body temperatures, even below freezing, with red bats (*Lasiurus borealis*) having been recorded as low as 23°F (−5°C), which implies the involvement of supercooling or cryoprotectants.

Hibernating bats arouse periodically, usually every two or three weeks, to move around and to drink the condensation from the cave walls and off their own fur. Several zoologists have remarked on sleeping bats defecating when handled, which appears to be proof of feeding, and caves also harbor hibernating insects, although not in sufficient number to sustain bats all winter. Many endothermic or warm-blooded animals with small body size and a very active lifestyle and high metabolism conserve energy by lowering their temperature every night during summer, irrespective of the air temperature, in the behavior known as daily hypothermia. Like the hummingbirds and the shrews, the northern bats also become torpid when asleep during the day, often with a considerable drop in temperature, perhaps as much as 36°F (20°C) below normal, and the heart beat and respiration of some bats sleeping during summer days is barely perceptible. Their daily summer sleep is very similar to winter hibernation, with their return to normal occurring quickly upon awakening and wriggling and shaking their wings. Some bats also become torpid daily during pregnancy and lactation, despite the assumed potential of lowered temperature for increasing the gestation length and reducing milk production. With their long winter hibernation and daytime sleep during the summer the northern insectivorous, nonmigratory bats therefore spend about three-quarters of their lives in a state of torpidity. Unfortunately, bats have suffered severely from the effects of human progress and development, and some species are endangered. Pollution, pesticides, colony disturbance; exclusion from buildings such as churches and cathedrals which were traditional roosts; treatment of wood with chemical preservatives, which are absorbed through the skin and have caused death or reproductive failure, have all seriously affected their numbers.

Some of the Species

Little Brown Bat (*Myotis lucifugus*)

The little brown bat is one of the many species of insectivorous *Myotis* bats, a tiny animal that weighs just ¼ ounce (7 g), and although simply brown it has a glossy sheen to the hairs on its back. One of its most distinguishing features is a long and pointed tragus—the leaf-like projection from the base of the external ear. It is a widespread species in North America, distributed from Alaska and central Canada south into the United States to a line drawn between California, Oklahoma, and Georgia. It is a very colonial bat, roosting in large numbers and in large clusters. It leaves its roost at dusk and returns at dawn, flying erratically as it catches insects on the wing, often using its tail membrane to create a pouch. Despite its small size the little brown bat has a very long lifespan, and up to thirty years has been recorded for banded wild specimens. It breeds in the fall, but development is delayed until February when fertilization occurs and embryonic growth begins. There is normally just a single baby, which is either carried by the mother or left hanging in the roost when she is away all night. From the far northern reaches of their range most individuals migrate to more southerly latitudes to hibernate, where a shorter period is required, and they favor caves, mine tunnels, buildings, or hollow trees for their long sleep, returning to their regular northern habitat in spring.

Greater Mouse-eared Bat (*Myotis myotis*)

The largest species of bat in Europe and Asia Minor, the greater mouse-eared bat has a wingspan of 16 inches (40 cm) and weighs 1.3 ounces (35 g). It has a dark sandy-colored upper body and paler underparts, large ears, and a prominent tragus. It is now very rare, near extinction in Great Britain, Holland, and Belgium, and increasingly rare elsewhere; it is considered Britain's rarest mammal. The increased use of agricultural pesticides is blamed for its disappearance, plus the loss of suitable roosting sites from forest clearance, mine closures, or disturbance. Many have also died from absorbing chemicals through continually hanging in buildings against wood treated with preservatives. The greater mouse-eared bat favors forests for feeding and roosts by day in tree cavities and buildings, especially church belfries. A solitary hibernator, it sleeps from October to March in central Europe, preferring the constant temperature and humidity of caves and deep rock crevices. It is a slow and straight flier, which feeds primarily on ground insects such as beetles, grasshoppers, spiders, and crickets, but it also catches moths in flight. Mouse-eared bats mate from August onward but fertilization of the egg is delayed and births do not occur until April or May. Females congregate in small colonies then and give birth to a single baby which is suckled for six weeks.

Eastern Red Bat (*Lasiurus borealis*)

The red bat is one of the common bats of North America, often called the tree bat because it is mostly a forest species that roosts in the open in trees—either on the trunk, among the leaves, or in the Southeast in Spanish moss. It is one of the more colorful species, the males having orange-red fur and the females grayish-chestnut, with the white tips to the hairs giving the coat a frosted appearance. It has short and rounded ears and its average weight is only ⅓ ounce (10 g). Its range is most of the United States, excluding the Rockies and the Great Basin, and southern

Eastern Red Bat   *A very hardy low-flying North American species, the eastern red bat can survive its body temperature dropping to 23°F (−5°C) during hibernation. It must therefore employ some form of freeze protection that protects the cell contents while allowing the extracellular fluids to freeze, as it is generally accepted that total freezing of the mammalian body is fatal.*
Photo: Merlyn D. Tuttle, Bat Conservation International

Canada up to the northwest shores of Hudson's Bay. A solitary species, it appears at twilight to hunt flying insects, especially beetles, low to the ground or beneath the forest canopy. This species varies in its wintering habits. Most of those that spend the summer in the northern end of their range migrate south to warmer latitudes for the winter, but the populations in the central United States are resident all winter and hibernate in tree hollows, caves, and sometimes in buildings. They have been observed flying out of piles of leaf litter, and are assumed to hibernate there also. Their long, silky fur provides some protection from the cold, and they are known to withstand body temperatures as low as 23°F (−5°C), which implies the ability to survive the freezing of their extracellular fluids. Females store sperm over winter and ovulate in spring when they emerge from hibernation; their offspring, usually twins, weigh only 1.5 g at birth.

Eastern Pipistrelle (*Pipistrellus subflavus*)

The pipistrelles are another large and widespread genus of small insectivorous bats, but there is considerable disagreement over their taxonomy. They are less colonial than the *Myotis* bats, and live in pairs or small groups. In North America the eastern pipistrelle is the common insect-eating species of the eastern half of the continent from southern Canada to northern Mexico. It is a tiny, dark yellowish-brown bat, weighing barely ⅙ ounce (4.8 g) and therefore sharing with Europe's common pipistrelle (*P. pipistrellus*) the title of the smallest mammalian hibernator. A slow and erratic flier, it seeks insects in open woods near water, catching them on the wing. It has small ears and a simple or plain nose that lacks a leaf adornment. Roosting singly or in small groups, it seeks dim places for daytime resting and for hibernating as it is very light-sensitive. From the northern-most latitudes of their range some individuals migrate south for the winter, but most stay and hibernate from October to March in caves, mine tunnels, rock crevices, trees, and building attics. It usually gives birth to two babies in the late spring or early summer, and carries them for the first week, after which they are then left hanging in the roost until they make their first flight at the age of four weeks.

Greater Horseshoe Bat (*Rhinolophus ferrum-equinum*)

The greater horseshoe bat has large pointed ears, small eyes, and a peculiar extension of the skin around its nostrils, called a nose leaf, which obstructs its vision. The lower part of this leaf is shaped like a horseshoe and surrounds the nostrils and covers the upper lip, while the upper part of the leaf is erect and pointed. It is associated with the emission of the bat's high-pitched calls as it flies with its mouth closed, and the sounds emitted through its nostrils are amplified and oriented by the leaf. It is a small bat, weighing just 1.3 ounces (35 g) and has a coat of silky-brown fur. Females have normal milk-secreting teats on the chest, plus two dummy (nonsecreting) teats on the lower abdomen to which the babies attach themselves with their teeth. It is a native of central and southern Europe where it prefers to hunt in woodland, fluttering like a butterfly on broad and rounded wings, hunting low to the ground and also alighting to seize insects. Whereas most small bats close their wings against their sides when they roost, the horseshoe bats wrap their wings around them and completely enclose their bodies, like the large flying foxes. In hibernation their body temperature drops from 104°F (40°C) to 46.4°F (8°C), and although they also hibernate in buildings and hollow trees, most individuals congregate in caves, clustered together for protection. In southern Russia, however, they hang singly from cave roofs and cold weather kills many in this exposed position, although they are also known to change hibernation sites in midwinter if their current one becomes unsuitable. Daytime sleep or daily torpidity is particularly deep in this species, and with low body temperature and barely discernable breathing and heart rate, resembles hibernation.

Pallid or Desert Bat (*Antrozous pallidus*)

This bat has desert coloration—creamy-brown upperparts with pale, almost white underparts—and has a small, horseshoe-shaped ridge on the muzzle over the nostrils. It weighs only 1 ounce (28 g), but has very large ears, which may be over 1 inch (2.5 cm) long, with a long and pointed tragus. The pallid bat is a native of western North America, from southern British Columbia through the western United States to Mexico, and also occurs on Cuba. It is a bat of the rocky desert scrub regions and up into the higher elevations of pine and oak forest; during the day, in the heat of the desert, it seeks the coolness of deep crevices, caves, and mine tunnels, which is where it also hibernates when food becomes scarce. Leaving its roost at dusk, it feeds for a few hours and then returns to the roost until just before dawn when it is active again. It has a low and swooping form of flight and seeks prey mostly on the ground, locating invertebrates such as beetles, moths, grass-hoppers, and even scorpions and small lizards with its extremely sensitive hearing. It swoops down, seizes the animal and carries it to a perch to eat. It also flits around flowering plants for the insects they may have attracted. The pallid bat is a social species that roosts in clusters, and is a late riser that leaves its roost in darkness. It mates in October but fertilization of the eggs within the female is delayed until March and the young are born in May, after an actual gestation period of just sixty days. Twins are normal in this species and the young are weaned at seven weeks and are sexually mature by October. In the northern parts of its range this species hibernates from October to March, this period being reduced to four months or even less in the south.

Note

1. Bats are occasionally seen flying during cold weather, generally as they move between roosting places.

9 Cold-Blooded Mammals

The division between the cold-blooded or poikilothermic fish, amphibians, and reptiles and the warm-blooded or homeothermic birds and mammals seems quite clear. The poikilotherms cannot generate their own heat and are totally dependent upon the environment to warm them or cool them. In total contrast the home-otherms have control of their thermoregulation, and produce or lose heat to maintain a constant body temperature within the set limits for the species, except when they become torpid during full-scale hibernation or short-term torpidity.

In reality, however, the situation is not as definite, for although the temperature of most mammals is quite high, averaging 98.6°F (37°C), in several groups thermo-regulation is poorly developed and their normal (nonhibernating) body temperature is much lower than the mammalian average. The echidnas' temperature fluctuates widely and averages 88.7°F (31.5°C), which is understandable in primitive mammals that retain so many reptilian characteristics; and the marsupials generally average 95°F (35°C). However, even lower body temperatures have been recorded for the tail-less tenrec and the two-toed sloth, whose active temperatures fluctuate with the environment, from 95°F (35°C) and 91.4°F (33°C), respectively, down to 75.2°F (24°C) in both species, making them the most cold-blooded of all the mammals.[1]

In addition to these species, which cannot maintain the high temperatures experienced by most mammals, there are also some that practice daily torpidity like many birds (see Chapter 6), a resting period with self-induced reduction of their body temperature and metabolic rate that reduces energy use. This normally occurs in response to food shortages or low temperature, but may happen at any time in species with high metabolism and large food requirements such as the shrews. This ability to become torpid on a daily basis has been reported in many mammals, especially the smaller marsupials.

These unusual mammals are included in this chapter, joined by other, more characteristic hibernators, such as the hedgehogs and some small, primitive primates.

■ SOME OF THE SPECIES

Monotremes

The first mammals to evolve from the reptiles in the Mesozoic Era about 185 million years ago were of two kinds, the Prototheria and the Theria. The latter were very successful and evolved into the modern live-bearing mammals, but the Prototherians were less successful and only the duck-billed platypus and echidnas survived to the present day. They are members of the order *Monotremata*, hence known as monotremes, and are the most primitive mammals alive, bearing characteristics of their reptilian ancestors and certain modern mammalian features. Unlike most placental mammals, they have a reptilian lack of thermoregulation[2] and their temperature therefore fluctuates with the environment. Like the reptiles and birds they have a cloaca, which is the single opening for passing fluid and solid wastes, for mating, and for laying eggs. They are the only mammals to lay eggs, which resemble those of the reptiles, with leathery skins and large yolks. Monotremes also resemble the reptiles in the structure of their eyes, and in certain aspects of their skulls, ribs, and vertebrae. Like the higher mammals, however, they have a coat of hair, or hair modified as spines, mammary glands for milk production, and they suckle their young; they also have a typical mammalian four-chambered heart, whereas most reptiles' hearts have three chambers. Young monotremes have teeth, but they lose them as they mature, so the adults are toothless and crush their food with their hard palates and tongues. At least one of the species—the short-beaked echidna—hibernates, and the duck-billed platypus may become torpid in cold water.

Short-beaked Echidna (*Tachyglossus ecaudatus*)

This monotreme resembles a giant hedgehog, with a compact muscular body covered with coarse hair and spines, supported on short, bandy legs and big feet with large claws. It has small ears and a stubby tail, and its snout is modified to form an elongated beak. It is one of the larger (noncarnivore) hibernators, reaching a weight of 13 pounds (6 kg), and is 14 inches (16 cm) long with spines about 2½ inches (6 cm) in length. The echidna is an Australian animal, which also occurs in southern New Guinea. It feeds mainly upon ants and termites, which are gathered with its long, sticky tongue, and then ground between the horny grooves on the back of the tongue and hard ridges on the palate. The echidna's small eyes are situated at the base of its snout and look ahead, providing binocular vision, although not very acute sight. Its olfactory sense is well developed and food is located by smell, aided by chemical-sensing vomeronasal organs, and possibly also with the help of the electroreceptors in the snout, which are believed to detect the electrical signals given off by its prey. The snout is a most remarkable organ, with receptors also for temperature and touch, and with great sensitivity to ground vibrations. Echidnas have good hearing, even though their external ears are almost hidden by hair.

The short-beaked echidna is primarily nocturnal and prefers cool temperatures, so it hides by day in a burrow or rock crevice and appears in the late afternoon. It also returns to its burrow at night to avoid high temperatures, as it suffers heat stress above 95°F (35°C). It lives in a wide range of habitat, including forest, grassland, and rocky areas, and when threatened either digs very quickly into soft soil, curls into a ball, or wedges itself into a crevice or burrow by erecting its spines. Both sexes have a spur inside their ankles, but unlike the male platypus, they lack venom glands.

Echidnas are solitary animals that socialize only at mating time. The single small egg is retained in the mother's body and provided with nutrients for some time before being laid. The pouch is a temporary one for the breeding season, and after laying the egg she maneuvers it into the pouch with her snout. The incubation period is only ten days, and like the premature young of the marsupials the tiny embryo is naked, blind, and helpless. It sucks very rich milk secreted by glands which open at the base of specialized mammary hairs (forerunners of mammary glands) onto two patches on the mother's belly. When the spines begin to sprout at eight weeks, she leaves her baby in a burrow and returns to feed it just once each week. At this stage of their lives, baby echidnas are very vulnerable to predators, especially feral cats.

The echidna cannot regulate its temperature as well as the metatherian (marsupial) or eutherian (placental) mammals, and although it is warm-blooded its body temperature fluctuates between 82.4°F (28°C) and 91.4°F (33°C) compared to the marsupial's 95°F (35°C) and the higher mammal's 98.6°F (37°C). It is therefore heterothermic[3] and could in fact be considered a "cold-blooded mammal," as its temperature fluctuates in accordance with the environment. As winter approaches and food supplies dwindle, echidnas cannot maintain a constant temperature and cool down quickly and become inactive, their body temperature dropping to 39.2°F (4°C) for extended periods, with just a single breath and heartbeat each minute, and they may cease breathing completely for several minutes. It is the only monotreme known to definitely hibernate, and in the colder parts of its range (Tasmania and Victoria) does so from late autumn to early spring, although animals that are reproductively active do emerge in winter. Their hibernation is broken by periodic arousals just as in the higher mammals.

Duck-billed Platypus (*Ornithorhynchus anatinus*)

The platypus is one of a kind, a single species in its own genus, an odd-looking animal, but a very successful one, for fossils show that it looks the same today as it did many millions of years ago. It has the streamlined body of an otter with a short, dense coat, composed of woolly underfur and blade-like guard hairs. It is blackish-brown with pale-yellow or brownish-white underparts, and has short, stout limbs and broad webbed feet, a beaver-like tail, and a leathery snout resembling a duck's bill. It lives in the rivers and ponds of eastern Australia, from Victoria to central Queensland, and in Tasmania, where it is active from dusk to dawn.

Ornithorhynchus is an excellent swimmer and diver and swims mainly with its forefeet. It cannot stay submerged for long, however, generally just a minute or two when feeding, but up to six minutes if it anchors itself—wedged beneath a submerged log, for example—to stop it bobbing up to the surface. It probes in the river and lake bottoms for aquatic invertebrates, tadpoles, small fish, freshwater crustaceans and snails, and holds the food in its cheek pouches until it returns to the surface to breathe. Horny plates in each jaw, ridged in front and smooth at the rear, crush the platypus's food. These plates are continually growing as they are worn down by grit scooped up with the prey.

Platypuses live in burrows on the banks of rivers and ponds, the entrance tunnel's diameter just wide enough to squeeze water from their coats before they reach the nest. They normally keep two burrows active, and their entrances are always above water level. One is for general living, for the pair; the other is used solely by the female as a maternity den. She lays two small eggs, the size of house sparrows' eggs, in a nest of damp leaves in her burrow. The incubation period is only ten days and she curls around them to provide the correct temperature. The tiny babies are then "brooded" and suckled by her for four months before they appear fully furred at the burrow's entrance. She plugs the entrance with soil while she is caring for her young.

Male platypuses are the only mammals able to actually inject[4] venom, through the hollow spurs on their ankles which are connected to poison glands. All baby platypuses have spurs but the females lose them as they mature. The venom can kill a dog but is not known to have caused death in humans. Their major predators are carpet pythons; the large monitor lizards called goannas; water rats, which kill the babies in the nest; and foxes and feral cats, which kill juveniles when they leave the nest burrow.

Platypuses lack outer ears or pinnae, so hearing is not their best sense. Underwater their ears and eyes are covered with folds of skin anyway, so they are effectively blind and deaf, and rely completely on their sense of touch. Their bills are extremely sensitive organs, which they sweep about to locate their food, and in addition to its tactile sense the bill also functions as an electroreceptor, detecting the weak electric fields emanating from small animals and locating them in the mud and under rocks. The special olfactory receptors known as vomeronasal organs, which are present in many reptiles, are well developed in the platypus. They contain sensory neurons similar to those in the nose that detect chemical compounds, including pheromones—which carry messages between animals of the same species—and are connected to the olfactory nerves.

Whereas the short-beaked echidna unquestionably hibernates, the case for the duck-billed platypus is unclear. David Fleay, considered the authority on the species for many years, stated that in Victoria it was dormant from late May to mid-September. Other reports have suggested that it may become torpid for up to a week at a time in midwinter (June to August in Australia). More recent accounts, however, insist that it remains active in winter in water at 32°F (0°C), well insulated by the air trapped in its underfur and between the outer guard hairs, and its only response to thermal stress is a reduction in weight and tail size. It is actually well suited for activity in cold weather as its normal body temperature is only 89.6°F (32°C).

Marsupials

The marsupials or metatherians are primitive mammals that occupy the position zoologically between the egg-laying monotremes and the higher placental or eutherian mammals. Although they are also called pouched mammals, the pouch or marsupium is very rudimentary in some species, being little more than a slight depression encircled by a ridge of skin. The common factor linking them all is their lack of a complete placenta—through which the higher mammals nourish their embryos in the womb—and marsupial young are therefore born after a very short gestation period. Their development continues in the pouch or in the absence of a pouch just dangling from a teat which swells in their mouth to firmly anchor them. Consequently, they are more correctly known as implacentals.[5]

Marsupials occur naturally in only two regions of the world; in the tropical New World, with just a single species ranging north into the United States; and in the Australasian Region—Australia, New Guinea, and their neighboring islands—which is their stronghold. Their evolution is entwined with the breakup of the southern supercontinent Gondwanaland about 100 million years ago, when Australia drifted away from South America with the early marsupials aboard before the evolution of the higher mammals, and especially the carnivores. The absence of the meat-eaters in early Australia gave the primitive, opossum-like marsupials the opportunity to adapt to many diverse niches, a process known as radiation. In doing so they independently developed similar characteristics to the developing mammals elsewhere in the world, which is called convergence or convergent evolution, their members filling the vacant niches with marsupials that eventually resembled the flying squirrels, mice, anteaters, cats, dogs, shrews, and even moles, of other continents. Like those animals they have evolved diets that are just as varied, including both terrestrial grazers and arboreal browsers, insectivores, carnivores and omnivores, and even scavengers.

Winter hibernation is not well represented in the marsupials, although many species practice daily torpidity. This is because most live in the world's warmer regions, with few species extending into the cold temperate zones beyond. In North America the only marsupial is the Virginian or common opossum (*Didelphys virginianus*) which ranges north to Minnesota and is therefore exposed to subfreezing temperatures. In South America the Patagonian opossum (*Lestodelphys halli*) has the most southerly distribution of the marsupials, living a terrestrial existence in the bleak grasslands of Argentina. The base of its tail thickens seasonally with stored fat, and it is therefore assumed to become torpid, but its habits are poorly known. Surprisingly, a central Brazilian species—the woolly opossum (*Caluromys lanatus*)—becomes lethargic when nights are cool, and its body temperature has dropped to 60.8°F (16°C) when the air temperature was 50°F (10°C), but the only South American species known to hibernate for long periods is the colocolo (*Dromiciops australis*).

Greater opportunities for marsupial hibernation exist at higher elevations where there is a wide seasonal temperature variation. Within the marsupial's range the mountainous regions are the Andes, the Alps of southeastern Australia, Cradle Mountain in Tasmania, and the central mountain ranges of New Guinea. Unfortunately, knowledge of marsupial biology in New Guinea is poorly documented, and of the Andean species is rather sparse, so several may well prove to practice torpidity. In the Chilean shrew opossum (*Rhynocholestes continentalis*) and some of the many species of mouse opossum (*Marmosa*), which range up to 13,000 feet (4,000 m) in the Andes of Boliva, Peru, and Chile, the tail thickens seasonally with stored fat. This is generally an indication of an animal preparing for lean times and lowered temperatures, and is usually associated with hibernation.

In Australia several species of pygmy possums[6] (*Burramyidae*) become torpid during the winter months, with the body temperature of the little pygmy possum (*Cercartetus lepidus*) cooling down to a level close to that of the ambient temperature. Although the actual length of its torpidity is unknown, the fact that it becomes very fat in body and tail implies a lengthy sleep. These tiny creatures weigh just ¼ ounce (7 g), and are the smallest mammalian hibernators after several bats and the birch mouse. The tiny feather-tailed glider (*Acrobates pygmaeus*), which weighs ½ ounce (14 g), also becomes torpid on cold days, and for several days at a time, therefore possibly qualifying as a hibernator and not a daily torpid

Little Pygmy Possum   *Its tail swollen with fat for its winter dormancy, this tiny marsupial from Australia and Tasmania, which weighs only ¼ ounce (7 g) without its fat stores, is one of the smallest terrestrial mammals to hibernate. It chills to within a few degrees of the ambient temperature during its long sleep.*
Photo: Dave Watts

species. However, length of torpidity is well documented for the slightly larger mountain pygmy possum (*Burramys parvus*), which lives high in the Victorian Alps and is currently considered Australia's only true seasonal hibernating marsupial.

Many other marsupials become dormant to avoid low temperatures, drought, and seasonal food shortages. Several of the native cats or dasyures do this, and the largest—the Tasmanian devil (*Sarcophilus harrisi*)—can become torpid at will, when its body temperature drops from 100°F (38°C) to 87°F (30.6°C), and its heart rate drops from 102 beats per minute to 85. Its respiration is also reduced, but it can return to normal in about one minute. The body temperature of the nocturnal cat-like eastern quoll (*Dasyurus viverrnus*) drops to 77°F (25°C) when it sleeps during the day. Sugar gliders (*Petaurus*), Australia's marsupial version of the placental northern flying squirrels, become torpid for a few days at a time when food is short or the temperature is low, and could therefore be considered hibernators. Several small marsupials living in the very arid "Red Centre" of the continent also undergo daily torpidity. They include the kultarr (*Antechinus laniger*), which has daily deep torpor in summer and shallower torpor in winter; the mulgara (*Dasycerus cristicauda*), which practices overnight torpidity and even becomes torpid when carrying young; and the fat-tailed pseudoantechinus (*Pseudantechinus macdonellensis*), a mouse-like marsupial that stores fat in its tail, which implies a long sleep. Several of the nocturnal marsupial mice or dunnarts (*Sminthopsis*) also practice daily torpidity, their temperature falling to 64.4°F (18°C) when they are asleep during the day.

Virginian or Common Opossum (*Didelphis virginianus*)

The only northern temperate zone marsupial, with a range from southeastern Canada and Minnesota to Central America, the Virginian opossum is the largest of the New World marsupials, with a head and body length of 20 inches (50 cm) when adult and a weight of 11 pounds (5 kg). It is nocturnal, both terrestrial and arboreal, and is a very agile climber, aided by its sharp claws and powerful prehensile tail, which is as long as the head and body and can support the animal's weight. Its unusual, sparse-looking coat is grayish-black with long white-tipped guard hairs, and it has a long and pointed grayish-white face, with bare black-and-white ears. "Possums" are solitary animals, best known for their habit of "playing possum" or feigning death, a state of catatonia in which they are totally immobile. It is uncertain whether this can be controlled by the opossum or is totally uncontrollable like fainting in humans. The opossum is an omnivore that hides in tree holes, burrows, or rock crevices during the day, and forages after dark for virtually anything edible, including invertebrates, amphibians, small reptiles, rodents, birds, and their eggs. It also eats carrion, fruit, and other plant matter, and in turn is relished by owls and by humans in some regions. Its well-developed pouch usually contains twelve nipples, and occasionally more, but there are rarely enough for all the babies that are born. When they find a teat they become attached as it swells in their mouth, so there is no sharing like piglets and puppies, and the excess young simply perish. The Virginian opossum is basically a tropical animal that has recently migrated north and may not have completely mastered the techniques of hibernation. In the colder parts of its range it increases its body weight by up to 30 percent with fat and becomes torpid for several weeks, usually sleeping in an abandoned woodchuck or skunk burrow, often communally. It frequently suffers frostbite damage to its ears, nose, and bare tail.

Colocolo or Monito del Monte (*Dromiciops australis*)

A small, mouse-like opossum, the only species in its genus, the colocolo is considered the sole survivor of a long-extinct family of marsupials, most closely related to those early opossums which were on the breakaway continent long ago and evolved into the Australian possums. It lives in the dense temperate rain forest of south-central Chile and neighboring Argentina, and on the island of Chiloe. It weighs only 1 ounce (28 g), has a head and body length of 4½ inches (11 cm), and a semi-prehensile tail of the same length, which is furred but has a bare strip on its underside. The colocolo has dense, short and silky brown fur, with white patches on the shoulders and rump, and pale underparts. It is mainly insectivorous but also eats soft fruit, and it makes a nest of waterproof leaves lined with grass and moss in dense vegetation, under a fallen tree or rock or in an elevated tree cavity. Females have a pouch and give birth to up to four tiny babies, after a gestation period of just twelve days. The colocolo practices daily torpor in winter, and hibernates for several weeks in the colder parts of its range when the temperature drops below 53.6°F (12°C); when preparing to sleep it can double its weight in about ten days, accumulating fat in the basal part of its tail to provide energy for its dormancy. It is a common animal within its restricted range, and is held in superstitious awe by the people of Chile's lakes region, who believe it is bad luck to have one indoors, and have burned down houses that became infested with them.

Fat-tailed Dunnart (*Sminthopsis crassicaudatus*)

A tiny, mouse-like marsupial, the dunnart lives in the arid inland regions of south and central Australia, in desert, steppe, dry woodland, and salt-bush country. There is considerable local variation in its pelage, from dark brown to ashy-gray and almost black dorsally, but with the underparts always creamy-white. It has large eyes and ears, a sharply pointed nose and a thick tail, and is 3½ inches (9 cm) long and weighs ½ ounce (14 g). The dunnart is highly carnivorous and is Australia's equivalent of the placental shrews, a very active animal with a high metabolic rate, which nightly consumes almost its weight in invertebrates, mice, and small lizards. It becomes torpid when food is short and when the temperature drops close to freezing, and in preparation for lean times it stores fat in its tail which enlarges to resemble a carrot. The dunnart normally breeds twice annually, giving birth to ten babies after a gestation period of only thirteen days, but often fails to reproduce when food is short. Young dunnarts remain in the pouch for six weeks and are then left in the nest for another three weeks, after which they become independent.

Mountain Pygmy Possum (*Burramys parvus*)

This marsupial was believed extinct for many years, after first being described from fossilized bones, but live specimens were discovered in 1956 on Mt. Higginbotham in Victoria's Alps, and then in 1996 on neighboring Mt. Buller. Although the largest of the pygmy possums, it is a tiny animal weighing only 1½ ounces (42 g), and has secretive nocturnal habits. It is the only truly alpine marsupial and the only marsupial to have a long seasonal hibernation, being torpid for almost seven months and surviving on just ⅔ ounce (20 g) of stored fat. It occupies an area of just 3 square miles (8 square km) at 4,265 feet (1,300 m) of elevation in mountain plum pine (*Podocarpus lawrencei*) heathlands, where it has to cope with alien predators such as foxes and feral cats and dogs, plus human activities including skiing. The pygmy possum is an omnivore that eats fruit, nectar, and insects, especially the bogong moth (*Agrotis infusa*). It is mainly terrestrial, and hibernation begins in March or April (the southern autumn) in a nest in a deep crevice or among rocks. Over winter the nest is covered with at least 3 feet 3 inches (1 m) of snow, which insulates the possums from the subfreezing temperatures of the mountains, the nest temperature remaining constant at 35°F (2°C). Torpidity lasts for about twenty days at a time, followed by arousal for about twenty-four hours when the possum may move around among the rocks beneath the snow. It sleeps until October or November, breeding when it emerges from hibernation, and the four babies stay in the pouch for only four weeks. They are then deposited in the nest until they are independent in January or February, when they must immediately begin to store fat in readiness for the forthcoming hibernation period. To supplement its fat reserves the pygmy possum also stores nonperishable foods such as seeds and roots, which it eats during its periodic arousals, the only marsupial known to do this.

Hedgehogs

During the evolution of the mammals most of the insectivores adopted nocturnal habits as protection from predators, but the hedgehogs have since evolved an additional protection in the form of a body covering of short, sharp spines. They can also roll into a ball to protect their vulnerable head, legs, and belly, and cannot be easily uncurled due to the contraction of their very muscular dorsal hood. They are animals of the Old World only, distributed across Eurasia and throughout Africa, living in wooded regions and deserts, hibernating in the north for the winter and estivating in arid regions to avoid drought and high summer temperatures. All hedgehogs are nocturnal and hide during daylight under piles of leaves or in holes which they dig with their powerful legs and strong claws. In the British Isles and Europe they live in wooded areas and farmland, generally in regions with rich soil and leaf litter, and do not usually burrow, but simply make a nest of grass and leaves. In Africa and central Asia they are desert animals; they dig burrows for their daytime rest and for hibernation and estivation, and can survive without water for up to ten weeks.

Hedgehogs are solitary and terrestrial, and although they are members of the primitive order *Insectivora*, in their eating habits they are carnivorous and consume a wide range of small animal life such as nestling birds, young mice and voles, worms and beetles, and are not averse to carrion. They are even cannibalistic, as caged animals have eaten their dead companions and wild hedgehogs feed on road-killed conspecifics. They have poor eyesight, but their hearing is good, and most species have medium-sized upstanding ears. Their sense of smell is also well developed and is used for finding food, for recognizing other hedgehogs, and for sensing danger; they locate earthworms and other soil invertebrates with their long and highly tactile snouts.

Spines are not good insulation and hedgehogs seem to feel the cold; since most live in regions experiencing cold winters, they have therefore evolved to hibernate. Those that occupy arid zones, which are intensely hot in summer and bitterly cold in winter, both hibernate and estivate. The three species of Eurasian hedgehogs (*Erinaceus*) occur across Europe and through central Asia to the Pacific Coast, including Siberia, Korea, and northeastern China, which have very severe winters, forcing the hedgehogs to hibernate for several months. Good eating in the fall prepares them for hibernation, and the European hedgehog (*Erinaceus europaeus*) makes its final nest preparations when the temperature drops to about 50°F (10°C); a further drop of 9°F (5°C) encourages it to roll tightly into a ball in the center of a nest of leaves and grass.

The seven species of desert hedgehogs of the genera *Paraechinus* and *Hemiechinus* range across North Africa and central and southwestern Asia, from the Ukraine to the Gobi Desert and south into Pakistan, India, and Iran, and must all escape seasonally harsh weather and lean times. In the desert regions of western India *Paraechinus micropus* and *Hemiechinus auritus* hibernate for at least three months, and then also estivate for several weeks during the hottest time of year, when food and water are scarce. In western Morocco the Algerian hedgehog (*Atelerix algerus*), one of the four species of African hedgehogs, hibernates in a burrow from October to March, and the very hot and dry summers force it underground again in June and July. Another African species, *Atelerix frontalis*, hibernates in South Africa between June and September, although it may emerge on warm days.

Long-eared Hedgehog (*Hemiechinus auritus*)

This species lives in the temperate semideserts and dry grasslands of central Asia from the Ukraine to Mongolia and south to northwest India, Pakistan, Iran, Asia Minor, and North Africa. True to its name it has long and flexible ears and long legs, its small spines or quills are lightly grooved and banded with brown and white, and it has a white belly. The average weight of an adult is 1 pound (454 g) and its head and body length is 9 inches (23 cm). The long-eared hedgehog's nocturnal and burrowing behavior and its ability to dispel body heat through its long, naked ears provide some protection from overheating in midsummer, but it may still estivate to escape intense heat. It hibernates from October to March in the

Long-eared Hedgehogs *An arid land species from central and southern Asia these hedgehogs hibernate for five months in the northern parts of their range, and for just three months in India's western deserts. They may estivate there also to escape the blistering heat.*
Photo: Courtesy I.V. Korneev, Leningrad Zoo

more northerly parts of its range to avoid low temperatures, usually burrowing beneath shrubs, and in the higher and colder elevations of Pakistan it is torpid for five months. In the Great Indian Desert of Rajasthan it hibernates for at least three months, and may also estivate during the hottest weeks of summer. Water is scarce in its environment, so it has little opportunity to drink and derives most of its fluids from its prey, but it is a very resistant animal that has survived without food or water for ten weeks in the laboratory. It breeds in spring immediately upon emergence from hibernation, and up to six young are born in a widened chamber at the end of its burrow after a gestation period of forty days.

European Hedgehog (*Erinaceus europaeus*)

A common animal in Europe, Sicily, Sardinia, and Corsica, this species is the most familiar hedgehog, although unfortunately most sightings are of roadkills. Its 1-inch-long spines are dark brown with yellowish-white tips, and usually number about 5,000; its face and underparts are grayish-brown. Adult hedgehogs are 12 inches (30 cm) long and have a tiny tail and short ears, and the average weight of an adult is about 2½ pounds (1.2 kg), but this increases by half with stored fat prior to hibernation. They roll into a tight ball to protect the head, legs, and belly, but as adults they have few predators other than the people who have traditionally eaten them. European hedgehogs prefer deciduous woodland, farmland, and scrub, and although terrestrial they can climb fences of wire mesh, and during daylight rest in old rabbit holes, among rocks, or under piles of brush or leaf mould. They hibernate there also—for up to five months in the most northerly parts of their range. One of their most unusual habits has been called anointing, when they lick objects excessively until they froth at the mouth, and the froth is then rubbed onto their spines. They have also been seen rubbing toads against their spines, similar behavior to "anting," where birds rub ants on their feathers, but the reason for this unusual behavior is unknown. The European hedgehog's body temperature is lower than that of most mammals, just 93.2°F (34°C), and it begins its long sleep when the ambient temperature drops to 41°F (5°C), at which time its own body temperature is close to that of the environment. The duration of hibernation in this species varies according to locale. In the British Isles it is seldom seen later than mid-November and reappears in March, whereas in the extreme south of its range, on the Mediterranean coast, it may not even hibernate. When dormant the hedgehog's heart beat plummets to 21 from the normal fast 188 per minute and breathing may cease for short periods. To prove this a hibernating hedgehog survived total submersion in water for twenty minutes, whereas an active (nonhibernating) animal would have drowned in a few minutes. Hedgehogs awaken every two or three days for several hours, and on warm nights may leave their hibernacula to feed. Increasing the temperature to 68°F (20°C) in the laboratory has awakened hibernating hedgehogs in several hours, but they did not become really active and search for food until their own temperature reached 93.2°F (34°C). Hedgehogs are able to suspend the start of their hibernation until conditions are right. Captive animals housed outdoors in England in the winter and provided with food regularly, did not hibernate unless the weather was very severe. Others remained active in the cold until provided with a nest box and material for nest-making.

Ethiopian Hedgehog (*Paraechinus aethiopicus*)

A desert hedgehog from North Africa and the Arabian Peninsula, the most distinctive feature of this species is the contrasting black muzzle and the broad white band across the face that extends back to separate the dark-brown spines from the ventral surface, which is very variable, but usually a mixture of black, brown, and white. Its long legs are dark and it has a bare patch on the forehead which divides the frontal spines. The Ethiopian hedgehog reaches a length of 8 inches (20 cm) and weighs about 17 ounces (500 g). It is at home in very barren regions, where it eats termites, beetles, locusts, and venomous snakes, although it is not immune, just very tolerant, to snake venom. During the day it shelters in rock crevices, under overhanging ledges, or in abandoned fox burrows, which are the favorite site for dormancy during cold weather, and it hibernates even in Arabia. Dormancy may last from November to late February, and when asleep its body temperature drops close to that of the environment. Up to four young are born in May or June after a gestation period of thirty-six days; they are deaf and blind and have short, soft spines even at birth.

Tenrecs

Tenrecs are the "hedgehogs" of Madagascar and its neighboring islands, and although they are not closely related they are remarkably like hedgehogs, with most species having spines, being tail-less, and some able to roll into a ball when threatened. They are descended from small mammals that became established on the islands just before it separated from Africa. In the absence of competitors they evolved to fill the vacant niches, including those occupied by hedgehogs elsewhere, and in doing so ended up resembling them, a case of convergent evolution. Like the hedgehogs, tenrecs are also members of the primitive[7] placental mammalian order *Insectivora*, and are currently divided into two subfamilies, the *Tenrecinae*, which contains the hedgehog-like spiny species; and the spineless and shrew-like members of the *Oryzoryctinae*. Their most interesting feature is their fecundity, with the tail-less tenrec (*Tenrec ecaudatus*) being the most prolific of all mammals, capable of producing up to thirty-two viable young per litter.[8] Like the reptiles and birds and the monotremes, tenrecs have a cloaca, which is the single urogenital and anal opening.

Tenrecs are terrestrial although they can climb. Most species are nocturnal, and they have good senses of smell and touch, but poor sight. Their hearing is believed to be acute, at least in the common tenrec, which makes rapid tongue-clicking sounds that may be a primitive form of orientation and prey location (like the bat's echolocation) for communication in the dark. Tenrecs are omnivorous and use their snout to dig for food. Prior to the start of the Austral winter (May to September) they eat more and lay down fat reserves within their bodies in order to hibernate, which they usually do in burrows with the entrance plugged with soil. The long-tailed or shrew tenrecs (*Microgale*) also store fat in their tails, and in Dobson's shrew tenrec (*Microgale dobsoni*) the normal weight of 1½ ounces (46 g) is almost doubled by the fat stored for hibernation. Madagascan winters are quite mild, and could be termed the cool, dry season rather than winter. In the highlands at 4,100 feet (1,250 m) the temperature averages 59°F (15°C) in the dry season, just a few degrees lower than summer levels, but the vegetation, and consequently the food supplies, suffer from lack of rain and the tenrecs become dormant in their burrows. Even captive tenrecs have followed this pattern and have therefore been allowed to hibernate for several months in accordance with their natural physiological rhythms, enabling them to reduce their accumulated fat.

Dormant tenrecs dug out of their burrows were cold to the touch, had a very low breathing rate, and had neither food in their stomach nor feces in the intestine. Even when active the tenrecs have a variable body temperature that ranges from 75.2°F (24°C) to 95°F (35°C). This is considerably lower than other mammals, which average 98.6°F (37°C), and the tenrec shares with the sloths the title of the most cold-blooded mammal. The body temperature of hibernating tenrecs is usually just 1.8°F (1°C) above the ambient temperature.

In addition to full hibernation the tenrecs also practice daily hypothermia, as they cannot maintain a constant body temperature, and cool down during their daily sleep like torpid hummingbirds. Members of the genera *Tenrec*, *Echinops*, *Setifer*, and *Hemicentetes* have become semi-torpid when the ambient temperature dropped below 68°F (20°C) and their own body temperature dropped by as much as 12.6°F (7°C), except in pregnant females and young animals. These tenrecs also experience a daily temperature range of several degrees between their active and rest periods during the spring and summer, and they also estivate during the hottest times of the year.

Tail-less Tenrec (*Tenrec ecaudatus*)

The tail-less tenrec occurs naturally on Madagascar, but has also been introduced and is now established on other western Indian Ocean islands. It lives in a variety of habitat wherever undergrowth provides hiding places, but is absent from the island's arid south. It has a stout body with pointed, hedgehog-like muzzle, short limbs, with the hindlimbs slightly longer than the forelimbs, and a head and body length of 1 foot 3 inches (38 cm) and a weight of 4 pounds 6 ounces (2 kg). Despite its name it has a tiny tail, just ¾ inch (2 cm) long. It has a grayish-brown or dark-brown coat of dense hairs and spines, which are often hidden by long guard hairs that can be erected on the nape. Long, tactile whiskers are probably this species' most important sensory adaptation, and it also has sensitive hairs on its back that detect vibrations. It has reasonably good eyesight, believed to be better than in the other tenrecs, and it scent-marks its territory to communicate with conspecifics, so it must also have a good sense of smell. It does not roll into a ball to protect itself.

Tenrec ecaudatus hibernates in underground burrows, plugging the entrance with soil, and is said to sleep continuously without rousing (which is very unusual for a hibernator) from May to October. Its normal active temperature varies greatly, from 75.2°F (24°C) to 95°F (35°C), but this has dropped to 55.4°F (13°C) during hibernation, with reduced heart rate and respiration. Specimens dug out of their burrows early in winter were excessively fat and were cold and sluggish with closed eyes and barely perceptible breathing. They paid little attention to their surroundings, and when left alone they resumed their sleep. None of the females removed from hibernation had embryos, yet individuals collected at the end of October were pregnant, which implies mating immediately upon emergence from their dormancy, sperm storage throughout hibernation, or delayed implantation. They are eaten by the natives, who particularly prize their fat deposits.

Pygmy Hedgehog Tenrec (*Echinops telfair*)

This is an arid-land species, a native of the dry southern tip of Madagascar, where it lives in forest and scrub, semidesert regions, and coastal sandy areas. It is a small, tail-less animal that reaches a length of just 7 inches (18 cm) and weighs

9 ounces (255 g); like the hedgehogs it can roll into a tight ball. It also resembles a hedgehog with its dorsal coat of short spines, except that they point in all directions. It has short legs and a hairy belly and face, and is variable in color, ranging from brownish-white to dark brown. The pygmy hedgehog tenrec is a good climber that sleeps, nests, and hibernates in tree holes and in ground burrows of its own digging. It is quite carnivorous and survives on invertebrates, young mice, and birds' eggs and nestlings. It has a captive longevity record of thirteen years, and may be the longest-lived tenrec; since it breeds readily it has been maintained as a laboratory animal in small cages for many years. *Echinops* hibernates during the southern winter, for periods varying between three and five months depending on the temperature. Mating occurs in October when it emerges from hibernation, and up to six young are born after a gestation period of sixty-three days.

Lowland Streaked Tenrec (*Hemicentetes semispinosus*)

The most distinctive tenrec, this species has long and hairy spines, which are barbed and detachable. It is blackish-brown with longitudinal chestnut stripes along its sides and back and one from the nose to the back of the head, and an erectable crest of spines on the nape. It is tail-less, reaches a length of 7½ inches (19 cm), and weighs up to 10 ounces (280 g). Although terrestrial, this tenrec is a good climber, and is predominantly an animal of the forests and secondary growth bushland where deep leaf litter supports earthworms, which form the bulk of its diet, and which it locates with its tactile muzzle. Single tenrecs make short tunnels with a nest chamber at the end, but family groups make more sophisticated burrow systems, which may be several yards long. Mothers and their babies communicate by vibrating special quills on their backs, which is known as stridulation. *Hemi-centetes semispinosus* is a less profound sleeper than the other *Tenrecinae* species, and some indiviuals may be partially active during the winter. Specimens caught in August and September were sluggish and cold after a cool night, but still had food in their stomachs so could not have been inactive for long. The temperature of hibernating streaked tenrecs was just 1.8°F (1°C) above the ambient temperature.

Shrews

Also members of the order *Insectivora*—the most primitive placental mammals— shrews are not hibernators. A long sleep is impossible for such active, highly strung animals with small body size and the need to a eat constantly to maintain their high metabolism. They are the mammalian equivalent of the hummingbird and could not possibly store sufficient fat to last a few days, let alone weeks or months. However, like the hummingbird and several other birds, plus the small insectiv-orous bats, this way of life is conducive to daily regulated torpidity—the lowering of their body temperature during their resting period as a means of conserving energy. It is a less profound state of inactivity than full hibernation, and occurs day or night, all year, or just seasonally; it is an animal's response to periods of food shortage or low temperature by lowering its metabolism and body temperature and therefore its demands for energy. The shrew's body temperature drops to about 59°F (15°C) when resting, from its normal 102°F (38.7°C). Shrews are the most numerous of the insectivores with almost 300 species. They are tiny, mouse-like animals, with beady eyes, tiny ears, and a very pointed muzzle. One of their members, the Etruscan shrew (*Suncus estruscus*), is one of the world's smallest mammals,[9] weighing just ⅟₁₅ ounce (2 g). Shrews range throughout the northern regions of both the Old and New Worlds and also occur in the mountains of northwestern South America. They prefer moist temperate regions where the mild climate and low risk of dehydration are important for their well-being, but they do live in regions experiencing harsh winters.

The shrews have several unique characteristics. They are the only mammals with a venomous bite; they are undoubtedly the most highly strung of all mammals, and some use a form of echolocation like the bats. They are solitary creatures that have a very short lifespan (occasionally three years, but usually less), but during that time can give birth to several litters averaging seven young per litter. Their small eyes are almost hidden by fur and they have very poor eyesight. Their senses of hearing and touch are well developed, but smell is their primary sense; and they have dermal glands that emit a strong scent that is used for territorial marking and for attracting a mate. In addition to the sounds they make that are discernable to the human ear as tiny mouse-like squeaks, several shrews also emit supersonic calls, which are reflected back from surrounding objects and possibly from potential prey, and function as an echolocation system, although nowhere near as sophisticated as that of the insectivorous bats.

Shrews have a very nervous disposition, and their rapid pace of life is powered by a normal heart rate of about 800 per minute, which can rise to well over 1,000 when they are severely stressed, which can be fatal. Their frantic lifestyle needs continual sustenance, so they are active both day and night, aggressively hunting a wide range of small animal prey, and resembling the mustelids (mink and weasels) in their predatory habits, rather than the other insectivores. They eat salamanders, small mice, nestling birds, worms, beetles, frogs, and the largest species can kill small rats; they normally eat their weight in food daily, but this rises to almost double their weight for pregnant females. Shrews, and the duck-billed platypus, are the only mammals able to inject[10] venom; the submaxillary glands of some shrews secrete a toxic saliva that acts as a nerve poison. With this venom they incapacitate earthworms and store them for later use, and they also store insects from which they have bitten off the legs to prevent them from escaping.

Although not related to the shrews, several species of elephant shrews (*Macroscelididae*) are also known to become torpid daily during periods of food shortages and low temperatures.

Common or European Shrew (*Sorex araneus*)

This shrew lives in Europe and northwestern Asia, and is one of the very common and widespread group of long-tailed shrews that occur across the northern regions of both the Old and New Worlds. It has a long and slender body and long, flexible snout, with a mouse-like tail, small eyes and small ears, and silky dark-brown coat with paler underparts. It is a tiny creature with a head and body length of about 3 inches (7.5 cm), a 2-inch-(5 cm)-long tail and weighs just ⅓ ounce (9.5 g). Although mainly nocturnal, it is also about during the day when hunting at night has not been too successful, and it is also active all year, making runways through the grass in summer and tunnelling beneath the snow in search of prey even in the depths of winter; it also burrows into loose soil. The common shrew is a highly carnivorous, solitary, and aggressive animal, which feeds voraciously on worms, insects, slugs, spiders, and carrion, probing and sniffing to locate food in the soil. It becomes torpid during its periods of rest, whenever these occur, with its body temperature dropping several degrees to lower its energy requirements. When a family is disturbed the babies follow their mother in line, each holding onto the tail of the one in front. Theses tiny animals are hunted by owls, weasels, stoats, and foxes.

Masked Shrew (*Sorex cinereus*)

Another member of the large genus of long-tailed shrews, the masked shrew is a common and widespread species in moist forest habitat across North America from the central United States northward into the Arctic Circle and in northeastern Siberia. It has a head and body length of 2 inches (5 cm) with a tail of 1½ inches (4 cm) and weighs about ⅓ ounce (10 g); it is grayish-brown above with paler underparts, and its tail is often bi-colored, brown and white. Like most shrews it can climb into low bushes. It breeds year-round in the southern parts of its range, but just in the summer months in the north, and is sexually mature at the age of four months. It is typically voracious and captive animals have eaten three times their body weight in food daily, while wild specimens are known to be active day and night in their constant search for food. More than 1,200 heartbeats per minute have been recorded for this species, but such stress-induced activity can be fatal.

Short-tailed Shrew (*Blairina brevicauda*)

A larger species from the eastern United States and southern Canada, the short-tailed shrew measures 3½ inches (9 cm) in length, but has a tail just 1 inch (2.5 cm) long. Adults weigh about ½ ounce (14 g) and have a slate-gray coat with paler underparts, and such tiny eyes they are barely visible. It has a wide range of habitat, including forests, marshes, and grassland, and is mainly terrestrial, but has also been seen in trees several feet above ground. The short-tailed shrew may be the only omnivorous species, for although it favors snails, beetles, and small mammals, it also eats nuts, berries, and seeds. It also stores food for later use, after immobilizing animate prey with its venom. It is more of a burrower than the other shrews and spends much of its life underground in a system of tunnels, which it digs itself with its strong claws and cartilaginous snout; but it also uses the burrows of other small mammals.

Sloths

The sloths are related to the armadillos and anteaters, which for many years have been collectively called edentates, after their order *Edentata*, which means toothless. However, the classification was clearly unsuitable as only the anteaters are toothless; the sloths and armadillos have continually growing premolars and molars, although no incisors or canines. Consequently, their order was recently renamed *Xenarthra*, based on their common characteristic of having articulations between the lumbar vertebrae which are known as xenarthrus vertebrae; but their long-used name is engrained and they will no doubt be called edentates for years to come.

The most well-known characteristics of the sloths are their sluggish movements and the fact that they spend practically all their lives hanging upside down, although this is only true of their active hours, for when resting they often prop themselves in a tree fork. They move slowly and definitely and may stay in the same tree for several days. Sloths are mainly nocturnal and are totally arboreal, and when they come down to the ground they crawl laboriously, but they do climb down to the base of their tree to defecate about once weekly. Their lazy appearance is deceptive as they can defend themselves by slashing with their forefeet and large, sharp claws, and by giving a powerful bite. They have ample teeth to cope with their leafy diet, but they are brown as they lack enamel, and their front premolars are pointed and resemble canine teeth. Sloths' eyes face forward and although their eyesight is poor they may have a degree of color vision. Their sense of smell is well developed but they have poor hearing, and their ears are barely visible beneath their long, shaggy hair. The sloth's arms are longer than its legs, so its head is always lower than its backside when it hangs from horizontal branches with its long claws, not its toes.

Sloths do not hibernate, but their body temperature is incompletely internally regulated and therefore fluctuates with the air temperature—an almost poikilothermic or cold-blooded physiology more akin to reptiles than mammals. With their low body temperature, and probably also due to low thyroid activity, the sloths have a low basal metabolic rate. They cannot control their body temperature due to their low muscle mass, and their body converts food to energy at about half the rate of other mammals. As they have not evolved any hibernation or migratory strategies for survival during lowered temperatures, the sloths are restricted to the lowland tropics.

Three-toed Sloth (*Bradypus tridactylus*)

This species differs in several ways from the two-toed sloth. It has three toes on all four limbs, and its forelimbs are slightly longer than the hindlimbs, whereas in the two-toed sloth they are considerably longer, and it has a coarse and shaggy gray coat, with a pale forehead and dark rings around its eyes. It is smaller, with a length of 20 inches (50 cm), and weighs about 10 pounds (4.5 kg). The three-toed sloth is also a more strict folivore, eating shoots and the young leaves of a variety of forest

Three-toed Sloth *Sloths are believed to have the least developed thermoregulation capacity of all mammals and they resemble reptiles in their uncontrolled hypothermia. The three-toed sloth's body temperature when active is usually about 93.2°F (34°C), several degrees below the mammalian average, but has dropped to 75.2°F (24°C) on cool days.*
Photo: Courtesy Stefan Laube

trees and little else, and the difficulty of replacing its normal diet outside its natural range has made it an unsuitable animal for zoological gardens. It has always been linked with the cecropia (*Cecropia*) tree which is a major source of food. To shed water the hairs of its coat point downward when it is hanging upside down from a branch. It has a small, rounded head with small eyes and ears, and it can turn its head through an arc of 270 degrees. Despite its already low temperature, which is usually below 93.2°F (34°C) during its nocturnal activities, the three-toed sloth's temperature when it is resting may drop by as much as 18°F (10°C), and on a dull day may be only 9°F (5°C) above the air temperature.

Two-toed Sloth (*Choloepus didactylus*)

The two-toed sloth lives in the tropical forest zone of South America from Colombia to Brazil. It has a head and body length of 2 feet 6 inches (76 cm) and weighs about 16 pounds (7.2 kg); it is grayish-brown with darker shoulders and a paler face, but frequently has a greenish tinge to its coat due to the presence of algae. Its forefeet have only two digits enclosed in skin right up to their claws, which are 3 inches (7.5 cm) long, but on the hind feet there are three sharp, hooked claws. The two-toed sloth's forelimbs are longer than the hindlimbs, it has a wide face with well-spaced and forward-pointing eyes; and individuals vary in their number of neck vertebrae—from six to eight. It is a more active animal than the three-toed sloth, and moves from tree to tree more often. Its single baby is born after a gestation period of about five months, and clings tightly to its mother's abdomen. Although it is a folivore this species has a wider-ranging diet than the three-toed sloth and also eats ripe fruit. Together with the tail-less tenrec the two-toed sloth has the most variable body temperature of all mammals, ranging from 75.2°F (24°C) to 91.4°F (33°C), depending upon the environment.

Primates

In the primates both hibernation and daily torpor occur only in several small lemurs, which are restricted to Madagascar and the neighboring Comoro Islands. Lemurs are prosimians—the most primitive primates—which are not at all monkey-like, and still resemble their ancestors that lived millions of years ago. Their actual origins, however, are unclear, although Africa is the most obvious source of their forebears. They were originally thought to have been on Madagascar when it separated from Africa about 160 million years ago, but it is now known that this separation happened long before the lemurs evolved. Their early primate ancestors must therefore have crossed from Africa, probably on floating vegetation, about 60 million years ago. The species that practice dormancy are all members of the family *Cheirogalidae,* the dwarf lemurs and the mouse lemurs. Five of the seven members of this family either hibernate (or estivate) or are torpid daily. They are the two mouse lemurs (*Microcebus*), the two species of dwarf lemurs (*Cheirogaleus*), and the hairy-eared dwarf lemur (*Allocebus trichotis*). All store fat in their thighs and their tails during the plentiful wet season (summer), and then become torpid for all or most of the dry season (winter), which lasts from May to October. They are all tiny animals, with the pygmy mouse lemur (*Microcebus myoxinus*) weighing just over 1 ounce (28.3 g) and therefore the smallest of the world's primates.[11]

To overcome the variations of food availability, these primates have evolved seasonal biological rhythms. They breed in the hot and wet summer and store fat then in preparation for their dormancy in the cool, dry winter when food is short. These seasonal rhythms are triggered by the lowered temperature and changes in the photoperiod. The term "hibernation" is normally used for winter dormancy when there is a marked seasonal temperature difference, and "estivation" for summer dormancy when shortage of food and water is the main reason for a long sleep. In Madagascar there is only about 18°F (10°C) difference between winter and summer temperatures, but the summers are wet and the winters are dry and food is scarce, so they become torpid then. Estivation therefore seems a more appropriate term for their long winter sleep.

Fat-tailed Dwarf Lemur (*Cheirogaleus medius*)

This species lives in the seasonally dry coastal forests of western and southern Madagascar. It has a soft and woolly, brownish-red or gray coat with white underparts and a white nasal stripe, and the large eyes of a nocturnal animal, encircled with dark rings; its head and body length is 8 inches (20 cm) and its tail is the same length. The fat-tailed dwarf lemur is very arboreal and seldom ventures to the ground. In preparation for the dry winter it stores fat in the base of its tail, and as a result its average weight of 5 ounces (140 g) increases to 7½ ounces (215 g). It

Fat-tailed Dwarf Lemur *In preparation for its long sleep during Madagascar's cool and dry winter (May to September), the nocturnal fat-tailed dwarf lemur lives up to its name and stores fat in its thighs and tail, increasing its weight of 5 ounce (140 g) by half. It estivates in a tree cavity nest with others of its kind.*
Photo: David Haring, Duke University

sleeps in a tree cavity together with others of its kind, rolled into a tight ball, from April to October. However, its temperature fluctuates several degrees depending upon the insulation value of its hibernacula. Males emerge from estivation first, to establish a territory for the family, and mating occurs when the females emerge and two or three young are born in January after a gestation period of sixty days. The fat-tailed dwarf lemur's main predators are owls and the fossa (*Cryptoprocta ferox*), a very agile, tree-climbing relative of the mongooses.

Hairy-eared Dwarf Lemur (*Allocebus trichotis*)

This is one of the rarest of the lemurs, believed extinct since 1875 until it was seen again in 1965. It is now thought to be restricted to a small area of rain forest in the vicinity of Mananara, on the eastern coast of Madagascar. It has a gray coat with a white stripe between the eyes, and a brown tail and brown ear tufts. It is one of the smallest primates, weighing just 3 ounces (85 g), and despite its size is said to be eaten by the local people; but apparently it has suffered most from deforestation. It lives in small family groups, and estivates from May to September, in a tree cavity or a ground burrow. It is nocturnal, arboreal, and very secretive, and little is known of its biology, but it is believed to mate in November, with the babies being born in January after a two-month gestation period, as well-developed young have been found in a tree cavity in March. This lemur is assumed to be naturally omnivorous, since captive animals ate soft fruit, honey, and insects.

Lesser or Grey Mouse Lemur (*Microcebus murinus*)

This lemur is one of the few plentiful species in Madagascar, due to its small size and unsuitability for eating. It lives in the coastal forests of western and southern Madagascar, both of which experience wet summers and dry winters. It has a thick and soft gray coat with a paler belly, a long, furred tail, large eyes that face forward for good stereoscopic (3D) vision, and thin and bare membranous ears for good sound reception. It weighs just 1½ ounces (42 g) and has a head and body length of 4½ inches (11.5 cm), just slightly larger than the pygmy mouse lemur (*M. myoxinus*). It does not appear to have a very effective thermoregulatory mechanism, and captive animals became semi-torpid and did not reproduce when kept at temperatures below 68°F (20°C). During the wet season in the wild they store fat in their hindlimbs and tail, which swells to 1½ inches (4 cm) in diameter, and then experience a decline in activity during the dry winter season, their periods of torpidity being triggered by the ambient temperature dropping below 64.4°F (18°C). Several huddle together in a tree hollow and are quite cool and stiff to the touch, their body temperature during estivation dropping below 68°F (20°C). They arouse through "burning" brown fat, a process known as non-shivering thermogenesis. They also practice daily torpidity, their body temperature dropping to 82.4°F (28°C) during the day and warming up at night to 95°F (35°C), which is their normal temperature.

Notes

1. The fossorial naked mole rat (*Heterocephalus glaber*)—a nonhibernator—is often said to be the most poorly thermoregulated mammal, but it has a body temperature of about 89.6°F (32°C).

2. A few placental mammals also have incomplete thermoregulation, and the sloths and tenrecs have even lower body temperatures than the monotremes.

3. In contrast, most mammals are homeothermic, able to maintain their body temperature irrespective of their environment.

4. The shrews have a mildly venomous saliva, which is forced into the bite wound when they chew on their victim.

5. Marsupials do have a very short-lived placenta, which contributes little to fetal development, so birth occurs after a brief gestation period.

6. The Australian marsupials were first called opossums, but this was changed to possum in the nineteenth century to avoid confusion with the New World opossums.

7. Considered primitive as they have a flat brain case and smooth brain, compared to the curved case and ridged brain of the primates. They also retain a cloaca, into which the digestive, urinary, and reproductive systems open, which is characteristic of the reptiles and early mammals.

8. The American opossum practices "super fecundity" and may give birth to more young, but cannot raise more than her total nipples, which is normally twelve.

9. It shares the title with the bumblebee bat (*Craseonycteris thonglongyai*) of Thailand, which also weighs $\frac{1}{15}$ ounce (2 g) when adult.

10. Shrews must chew to break the skin and allow the venom to enter the wound, like the rear-fanged venomous snakes, whereas the platypus injects venom via the spurs on its hind legs.

11. The pygmy marmoset, so often called the world's smallest "monkey," is considerably heavier, at 4.2 ounces (119 g).

Glossary

Adaptive radiation
The diversification of a species as it adapts to different ecological niches and eventually becomes so specialized for the new environment that it evolves into a different species.

Adipose tissue
Specialized connective tissue, usually found in a layer beneath the skin, which can store large amounts of neutral fats. These cushion the body's organs and serve as a source of energy.

Amnion
The innermost thin membrane forming a sac that surrounds the fetus, and containing amniotic fluid—the clear liquid that protects the fetus and allows it to exercise. The chorion is the outer membrane.

Anabolism
One of the two major categories of biochemical processes involved in metabolism. A metabolic reaction that builds up complex molecules, called the "construction phase" of metabolism. The opposite of catabolism.

Apnea
The temporary suspension of breathing, which occurs in many hibernating animals, often for periods of hours and occasionally days, in conjunction with low heart rate and temperature.

Articulation
The point of connection between two bones or elements of a skeleton, so that motion is possible.

Autotomy
The voluntary shedding of a body part, such as a lizard's tail, usually in defense, and the ability to regenerate a new one.

Basal metabolic rate (BMR)
A measurement of energy—the minimum calorific requirements needed to sustain the basic functions of life such as breathing and blood circulation when the body is at rest.

Biosynthesis
The synthesis (production) of essential chemical compounds by living cells (usually from more simple molecules). An essential part of anabolism.

Brown fat
Dark-colored fatty tissue in newborn and hibernating mammals, which is rich in mitochondria, and generates heat to regulate body temperature. Used by hibernating animals to warm up during their arousal.

Carapace
The top shell of a turtle or tortoise.

Carnivorous
Having a diet consisting mainly of the flesh of other vertebrates—meat, fish, shellfish, snakes, frogs, etc. When members of the same species are eaten it is considered cannibalism.

Catabolism
One of the categories of biochemical processes involved in metabolism. The breakdown of the complex molecules built up by anabolism and the release of their energy.

Catatonia
Immobility with muscular rigidity and inflexibility, as seen in animals like the Virginian opossum and the hog-nosed snake, which "play dead" as a safety measure.

Circadian rhythm
A "daily cycle" prompted by an "internal clock." The repetition of certain events in living creatures at the same time within each period of twenty-four hours, which includes eating and sleeping.

Circannual rhythm
A biological event occurring at approximately the same time each year, such as the commencement of hibernation or migration, prompted by an "internal clock."

Cochlea
The snail-shaped, fluid-filled cavity in the inner ear that converts sound vibrations from the middle ear into nerve impulses which travel to the brain.

Convergent evolution
The development of similar structures in distantly related animals as a result of adapting to similar environments or life strategies. The evolution of the tenrecs in Madagascar, for example, has resulted in species that not only physically resemble hedgehogs but also hibernate and estivate like them.

Cryoprotectant
Anti-freeze, produced by some northern animals to prevent the freezing of their intracellular fluids.

Daily torpidity
Semilethargic, energy-saving sleep of a few hours' duration, by day or night and at any time of year.

Dermal
Referring to the skin. The layer of skin below the surface layer which is called the epidermis. Dermal absorption of oxygen and water occurs through the skin of the amphibians.

Devonian Period
The "Age of Fishes" from 410–360 million years ago, during which time the first tetrapods or land-living vertebrates appeared.

Diurnal
Active during the hours of daylight.

Dorsolateral
Pertaining to an animal's sides and back.

Ecdysis
The periodic shedding of the exoskeleton by insects and reptiles to allow growth.

Ectotherm
A cold-blooded or poikilothermic animal, dependent for its body heat on the environment.

Ectothermy
Temperature regulation in animals, which depends primarily upon the absorption of heat from the environment. Antonym of endothermy.

Endemic
Native to or confined to a certain region.

Endocrine glands
Glands that produce and secrete hormones into the bloodstream, which regulate the body's normal functions.

Endotherm
An animal (bird or mammal) capable of maintaining its body temperature.

Endothermy
Thermoregulation, which depends mainly upon a high and controlled rate of heat production by metabolism and the dissipation of the excess to the environment. Antonym of ectothermy.

Estivation
Long-term summer sleep to avoid hot and dry weather and to conserve water and energy, while surviving upon body fat.

Eutheria
The placental mammals, all species above the monotremes and marsupials.

Euthermia
The normal warm-blooded state of mammals and birds.

Extracellular
Outside the cell. The noncellular body fluids—blood, urine, etc.—which freeze in some northern amphibians when the contents of their cells are protected by cryoprotectants.

Feral
A domesticated animal that has successfully returned to the wild.

Freezing
The withdrawal of heat from an animal's body that changes liquids into solids. The point at which a liquid becomes a solid.

Frostbite
Injury or destruction of skin and tissue, usually of the extremities, due to prolonged exposure to freezing temperatures.

Genus
A biological classification. A grouping of living organisms having one or more related or morphologically similar species. The plural is genera.

Glucose
The simple sugar that is the animal body's main source of energy.

Glycerol
A syrupy substance obtained from the fatty acids of fats.

Glycogen
The main form of carbohydrate energy that is stored in the liver and muscles, and which releases glucose into the blood when it is required by the cells.

Glycogenolysis
The breakdown of glycogen to glucose.

Glycolysis
The process in which glucose is broken down by the cells to produce energy in enzyme reactions which do not need oxygen.

Granular
Having a grainy surface.

Heliotherm
An ectotherm that derives its body heat from the sun.

Herbivorous
Having a diet consisting almost solely of plant material, but generally of leaves and therefore either grazers or browsers.

Heterotherm
An animal that has a fluctuating but controlled body temperature, such as the mammals that hibernate.

Heterothermy
Thermoregulation in birds and mammals, in which there is a periodic major variation in the core temperature.

Hibernation
Any form of long-term torpidity in ectotherms and endotherms to escape unfavorable winter conditions (of climate and food supply) and to conserve energy, while surviving upon body fat or external food stores.

Holarctic
A zoogeographic region encompassing the temperate zones of the Northern Hemisphere.

Homeotherm
An organism (a mammal or bird) capable of maintaining a stable body temperature irrespective of the environment.

Homeothermy
Thermoregulation within a narrow "set range" in birds and mammals, with a high level of basal metabolism characterized by small variations of body temperature, usually plus or minus 3.6°F (2°C).

Hypomelanistic
A mutation in which the black and brown pigment is absent or reduced.

Hypothermia
Dangerous loss of body warmth resulting in abnormally low body temperature, through losing heat faster than it is produced. Potentially lethal to man and all mammals and birds except the hibernating species; causes respiratory failure and cardiac arrest.

Hypoxic
Insufficient oxygen in an environment to support aerobic respiration, such as the mud at the bottom of ponds frozen over in winter.

Indigenous
Native to a particular country or area.

Insectivorous
In theory, having a diet consisting mainly of insects, but in practice most insectivorous animals eat invertebrates (worms, snails, spiders, etc.) in addition to insects, and are also likely to eat small vertebrates such as lizards, snakes, and frogs, and are therefore partially carnivorous.

Intracellular
The contents of a cell, its cytoplasm or protoplasm and the nucleus.

Jacobsen's organ
An extrasensory organ in the roof of the mouth of many animals, which receives scent particles carried by the air or on the tongue and transmits the information to the brain for action. Also known as the vomeronasal organ.

Keratin
The major protein component of hair, wool, horn, hoofs, nails, and feather quills.

Mammary gland
The glands present in female mammals that produce milk for suckling the young. They are believed to have evolved from sweat glands.

Marsupium
The pouch in which the prematurely born and helpless young of the kangaroos, possums, and related species of marsupials are raised.

Maxillary
Relating to the upper jawbone.

Metabolic depression
An animal's response to hard times, especially lack of food, by lowering its resting metabolic rate to limit energy expenditures.

Metabolic rate
The rate at which an organism transforms food into energy and body tissue. The amount of energy liberated per unit of time.

Metabolism
The chemical processes within the body that sustain life. Some substances are broken down from ingested foods to provide energy, others are synthesized internally by the animal. Two main processes are involved: anabolism, which is the biosynthesis of complex organic substances from simpler ones, and catabolism, which is the breakdown of complex substances to release their energy.

Metamorphosis
The process of changing from one form to another, such as the transformation of a tadpole (a larva) into a froglet.

Metatarsal tubercle
The growth or "spade" on the hind feet of certain toads to aid their burrowing into soil.

Metatheria
The marsupials, primitive implacental mammals between the Prototheria (the monotremes) and the Eutheria (the placental mammals).

Mitochondria
Structures in a cell's cytoplasm that contain enzymes for cell metabolism, especially for converting food into energy.

Monotypic
Having only one representative, such as a genus with a single species.

Morphology
The form or structure of an organism or any of its parts.

Mutation
Any change in the DNA (deoxyribonucleic acid) of a cell, which can be inherited if it occurs in cells that produce sperm or eggs, and results in an animal that differs from its parents in some manner, although this may not always be obvious.

Nearctic
The zoogeographic region encompassing the Arctic and temperate regions of the New World—North America south to central Mexico.

Nocturnal
Active during the hours of darkness.

Normothermic
The condition of normal body temperature.

Nucleus
The part of a cell containing DNA (deoxyribonucleic acid) and RNA (ribonucleic acid), which are responsible for growth and reproduction.

Omnivorous
Having a varied diet that contains foods of plant and animal origin.

Ossified
Hardened into bone by the deposition of calcium.

Osteoderms
Bone embedded into the skin, such as the small bony plates in the glass lizard's scales to reinforce them and create rigidity.

Ovulation
The release of a mature egg from the surface of the ovary.

Palaearctic
The zoogeographic region encompassing the Arctic and temperate zone of the Old World—Europe and Asia and Africa north of the tropics.

Palate
The upper surface of the mouth that separates the oral and nasal cavities.

Parotid glands
Salivary glands located in the mouth near the ears.

Parotoid glands
Large skin glands situated on the neck, sides of the head, or shoulders of many species of toads, which secrete a highly toxic milky substance.

Patagium
The flap of skin along the sides of the body used for gliding by the flying squirrels and sugar gliders, and the wing used for true flight by the bats.

Permian Period
The last period of the Paleozoic Era, from 290 to 248 million years ago, which ended with the greatest mass extinction of animals in earth's history, when more than 95 percent of all species disappeared from the fossil record.

Phalanges
The bones of the fingers and toes.

Plastron
The lower shell of a tortoise or turtle.

Poikilotherm
A cold-blooded animal whose body temperature varies in accordance with its surroundings, therefore all the vertebrates except birds and mammals.

Polytypic
Having several variant forms, especially a species with a number of subspecies or races.

Prototheria
The monotremes (the echidnas and the duck-billed platypus), which are the most primitive of the mammals and still retain several reptilian characteristics, including laying eggs.

Quadrate bone
A bone in birds and reptiles that articulates the lower jaw with the skull.

Saltatorial
Having a jumping or bounding form of locomotion as in the jerboas and kangaroos.

Sloughing
The shedding of its dead outer skin by a reptile or amphibian.

Species
A group of related organisms that are capable of interbreeding and of producing fertile offspring.

Spermatogenesis
The process of the formation of spermatozoa (sperm).

Stereoscopic vision
Synonymous with binocular vision, in which forward-facing eyes produce a single, three-dimensional image in the area of overlap of the visual fields, which provides increased depth perception.

Subspecies
A taxonomic subdivision of a species, capable of interbreeding, and synonymous with "race."

Supraocular
Scales above the eye in lizards and snakes.

Taxonomy
The scientific naming and classification of living organisms (plants and animals), begun by Swedish biologist Carl Linnaeus (1707–1778). It is one of the three branches of Systematics, the others being Identification (the animal's description) and Phylogenetics (the animal's relationship to others).

Tetrapod
A vertebrate with four limbs.

Thermogenesis
The production of heat in warm-blooded animals by increasing the metabolic rate and the breaking down of fat molecules.

Thermoregulation
The maintenance or regulation of normal body temperature in mammals and birds.

Torpidity
A state of motor and mental inactivity in varying degrees, from the cold and "lifeless" profound long-term hibernators such as the ground squirrels and the poorwill, to the species that practice daily torpor with lowered temperature like the shrews and hummingbirds.

Triassic Period
The first phase of the Mesozoic Era, called "The Age of the Dinosaurs," from 248 to 208 million years ago.

Tympanic membrane
The eardrum or tympanum, a thin membrane that separates the outer ear from the middle ear, and vibrates in response to sound waves. It lies exposed as a round spot behind the eye in the frogs and toads.

Urea
The chief solid waste component of animal urine, which is excreted in the urine.

Uric acid
A substance in the blood and urine that is a by-product of protein metabolism.

Vasoconstriction
Constriction of the blood vessels and decrease of local blood flow.

Bibliography

Animal Diversity Web. http://animaldiversity.ummz.umich.edu/site.

Anon. Endangered Species. www.redlist.org/.

Ask the Experts: Biology. http://www.sciam.com/askexpert_question.

Bailey, V. *Mammals of the S.W. USA.* Dover, New York, 1971.

Banfield, A.W.F. *The Mammals of Canada.* University of Toronto Press, Toronto, 1974.

Barbour, R. W. and Davis, W. H. *Bats of America.* University Press of Kentucky, Lexington, 1969.

Bats. http://www.batcon.org/.

Biology Dictionary. www.biology-online.org/dictionary.

Broome, L. S. and Geiser, F. Hibernation in free-living Mountain Pygmy Possums (*Burramys parvus*) (Marsupialia: Burramyidae). *Aust. Jour. Zool.* 43 (1995): 373–379.

Brown, L. and Amadon, D. *Eagles, Hawks and Falcons of the World.* Country Life Books, Wisbech, Cambridge, 1968.

Bucher, T. L. and Worthington, A. Nocturnal hypothermia and oxygen consumption in manakins. *Condor* 84 (1982): 327–331.

Burt, W. H. and Grossenheider, R. P. *A Field Guide to the Mammals.* Houghton Mifflin, Boston, 1964.

Conant, R. *A Field Guide to the Reptiles and Amphibians of Eastern/Central North America.* Houghton Mifflin, Boston, 1975.

Costanzo, J. P. and Claussen, D. L. Natural Freeze Tolerance in the terrestrial turtle Terrapene ornata. *J. Exp. Zool.* 254 (1990): 228–232.

Definitions. http://dict.die.net/.

Devis, D. E. Hibernation and Circannual Rhythms of food consumption in marmots and ground squirrels. *Quarterly Review of Biology* 51(4) (1976): 477–514.

Eisenberg, J. F. The Heteromyid Rodents. In *The UFAW Handbook on the Care and Management of Laboratory Animals.* Livingstone, Edinburgh and London, 1966, 391–395.

Eisenberg, J. F. and Gould, E. The maintenance of tenrecoid insectivores in captivity. *Int. Zoo Yrbk.* 7 (1967): 194–196.

Endangered Species. ARKIVE. www.arkive.org/species.

Geiser, F. Hibernation: Endotherms. In *Encylcopedia of Life Sciences.* Macmillan Publishers Ltd., Nature Publishing Group, London, 2001.

Grigg, G., Beard, L., Grant, T., and Augee, M. Body temperature and diurnal activity patterns in the Platypus (Ornithorynchus anatinus) during winter. *Austr. J. Zool.* 40 (1992): 135–142.

Harcourt, C. and Thornback, J. *Lemurs of Madagascar and the Comoros.* The IUCN Red Data Book. IUCN, Gland, Switzerland and Cambridge, UK, 1990.

Heldmaier, G. and Klingenspor, M. (Eds.). *Life in the Cold.* The 11th International Hibernation Symposium. Springer Verlag, New York, 2000.

Hibernation. http://whyfiles.org/187hibernate.

Hudson, J. W. and Wang, L.C.H. *Strategies in Cold.* Academic Press, New York, 1978.

Jaeger, E. C. Does the Poor-will "hibernate"? *Condor* 50 (1) (1948): 45–46.

Jaeger, E. C. Further observations on the hibernation of the Poor-will. *Condor* 51 (3) (1949): 105–109.

Kayser, C. *The Physiology of Hibernation.* Pergamon Press, Oxford, 1961.

Lasiewski, R. C. The energetics of migrating hummingbirds. *Condor* 64 (1962): 324.

Lehmer, E. M., Van Horne, B., Kulbartz, B., and Florant, G. L. Facultative Torpor in Free Ranging Black-tailed Prairie Dogs (Cynomis ludovicianus). *J. Mammal.* 82 (2) (2001): 551–557.

Lioncrusher's Domain. http://www.lioncrusher.com.

Lowery, G. H. Jr. *The Mammals of Louisiana and its adjacent waters.* Louisiana State University Press, Baton Rouge, 1974.

Lyman, C. P. Activity, food consumption and hoarding in hibernators. *J. Mammal.* 35 (1954): 545–552.

Lyman, C. P., Willis, J. S., Malan, A., and Wang, L.C.H. *Hibernation and Torpor in Mammals and Birds.* Academic Press, New York, 1982.

Mayer, W. V. *Hibernation.* D. C. Heath, Boston, 1964.

Melissa Kaplan's Herp Care Collection. http://www.anapsid.org.

Morris, B. The European Hedgehog. In *The UFAW Handbook on the Care and Management of Laboratory Animals.* Livingstone, Edinburgh and London, 1966, 478–488.

MSN Encarta (species). http://ca.encarta.msn.com/encyclopedia.

Nowak, R. M. *Walker's Mammals of the World.* Vols. I & II. Johns Hopkins University Press, Baltimore and London, 1991.

Ognev, S. I. *Mammals of Eastern Europe and Northern Asia.* Israel Program for Scientific Translations, Jerusalem. 8 vols. 1962–1964.

Pearson, O. P. The Metabolism of Hummingbirds. *Condor* 52 (1950): 145–152.

Pinder. A. W., Storey, K. B., and Ultsch, G. R. Estivation and Hibernation. In *Environmental Physiology of the Amphibians.* Edited by M. E. Feder and W. W. Burggren. University of Chicago Press, Chicago, 1992, 250–274.

Rand, A. L. On the habits of some Madagascar animals. *J. Mamm.* 16 (2) (1935): 89–104.

Ride, W.D.L. *A Guide to the Native Mammals of Australia.* Oxford University Press, London, 1970.

Rogers, L. A Bear in its Lair. *Natural History* 90 (10) (1981): 64–70.

Roots, C. Manitoba's Great Snake Harvest. *Country Life* (February 1985).

Russian Bats. http://www.zmmu.msu.ru/bats.rbgrhp/autjor1.html.

Schmid, W. D. Survival of frogs in low temperatures. *Science* 215 (1982): 697–698.

Stebbings, R. E. *Conservation of European Bats.* Helm, London, 1988.

Stebbins, R. C. *A Field Guide to Western Reptiles and Amphibians.* Houghton Mifflin, Boston, 1960.

Stirling, I. *Polar Bears.* University of Michigan Press, Ann Arbor, 1988.

Storey, K. B. Life in the Slow Lane: molecular mechanisms of estivation. *Comp. Biochem. Physiol.* A133 (2002): 733–754.

Storey, K. B. and Storey, J. M. Freeze tolerant frogs. *Can. J. Zool.* 64 (1986): 49–56.

Storey, K. B. and Storey, J. M. Natural Freezing Survival in Animals. *Ann. Rev. Ecol. Syst.* 271 (1996): 365–386.

The Carey Lab. http://www.vetmed.wisc.edu/cbs/carey2.

Tortoises. www.tortoisetrust.org.

Turtles. www.austinsturtlepage.com.

Index

About the Author

CLIVE ROOTS has been a zoo director for many years. He has travelled the world collecting live animals for zoo conservation programs. Roots has acted as a masterplanning and design consultant for numerous zoological gardens and related projects around the world, and has written many books on zoo and natural history subjects.